Disaster Robotics

Intelligent Robotics and Autonomous Agents
Edited by Ronald C. Arkin

For a complete list of the books published in this series, please see the back of this book.

Disaster Robotics

Robin R. Murphy

The MIT Press
Cambridge, Massachusetts
London, England

MIT Press books may be purchased at special quantity discounts for business or sales promotional use. For information, please email special_sales@mitpress.mit.edu.

This book was set in Stone by the MIT Press. Printed and bound in the United States of America.

Library of Congress Cataloging-in-Publication Data
Murphy, Robin, 1957–
Disaster robotics / Robin R. Murphy.
   pages   cm
Includes bibliographical references and index.
ISBN 978-0-262-02735-9 (hardcover : alk. paper)
1. Robots in search and rescue operations. 2. Mobile robots. I. Title.
TJ211.46.M87 2014
363.34—dc23
2013032411

10 9 8 7 6 5 4 3 2 1

To Kevin, Kate, and Allan

Contents

Preface   ix

1   Introduction   1
2   Known Deployments and Performance   21
3   Unmanned Ground Vehicles   63
4   Unmanned Aerial Vehicles   111
5   Unmanned Marine Vehicles   137
6   Conducting Fieldwork   163

Common Acronyms and Abbreviations   197
Glossary   199
References   207
Index   217

Preface

The purpose of this book is to provide a comprehensive resource for researchers, practitioners, and the public on the use of robots for any type of disaster. The recent series of earthquakes, tsunamis, oil spills, and mine disasters has heightened the awareness of the value of robots for response and recovery as well as for prevention and preparedness. The book is intended to serve as an *introduction* for researchers and technologists who need to understand the domain and the state of the art to apply their research and development expertise to this emerging field; as a *reference manual* for agencies and emergency managers on the state of the practice; and as a *textbook* for students in field robotics and human–robot interaction. The book also acts as a *history* of disaster robotics for the public and science writers seeking to place current technologies in context with past innovations or deployments and to project future uses and challenges.

Motivation

The motivation for writing this book stems from my engagement with disaster robotics, now approaching 20 years. My primary goal is to pull together the objective data and published findings about disaster robotics from all sources in more detail than captured in the chapter on search and rescue robotics that I co-wrote with many colleagues for the *Handbook of Robotics*. My secondary goal is to share my "up close and personal" experiences from the field. As a field researcher, I have adopted many methods from anthropology and cognitive science, especially the technique of reporting case studies. Though my experiences do not support statistical inference, the case studies offer a narrative on the complex sociotechnical ecology within which robots must function. I also hope that the case studies impart at least some of the sense of excitement and discovery that has made my career so rewarding.

My involvement in disaster robotics started in 1995, when my research shifted from artificial intelligence for mobile robots in general to artificial intelligence specifically for disaster robots. The shift came in the aftermath of the twin disasters of that year: the bombing of the Alfred P. Murrah Federal Building in Oklahoma City in the United States and the Kobe earthquake in Japan. One of my graduate students at the Colorado School of Mines, John Blitch, then a U.S. Army major with search and rescue training, participated in the response at the Oklahoma City bombing. His experiences motivated his master's thesis that combined artificial intelligence and operations research to analyze the utility of rescue robots, and they were also responsible for my conversion to disaster robots.

My disaster robotics research shifted from the laboratory to the field in 1999 after my move to the University of South Florida. At the University of South Florida, my group began working with the Hillsborough County Fire Rescue Department and Florida Task Force 3. Under the tutelage of Chief Ron Rogers, we had to discard most of our laboratory-based research in artificial intelligence and marsupial platform designs where a mother robot would carry and deposit a daughter robot. The research was not addressing the needs or the realities of fire rescue operations. I drew on my fond memories of an anthropology course that I took as an undergraduate and started conducting ethnographies during exercises, creating the foundation for my work in human–robot interaction. We began routinely participating in exercises and even taking response classes to learn more about the realities of disaster response. The fieldwork changed not only our research but also the type of equipment we purchased with grants, how we packed and stored the gear for ease of transportation to field exercises, the tools and spare parts we brought along, and the data we learned to collect.

The shift from field research to actual deployments started in 2001. On September 11, 2001, my group joined the Center for Robot-Assisted Search and Rescue (CRASAR) for the response to the collapse of the World Trade Center, the first reported use of rescue robots. CRASAR had been formed 2 weeks earlier by John Blitch. He was in the process of stepping down as the manager of the Defense Advanced Research Projects Agency (DARPA) Tactical Mobile Robots program, which created the iRobot Packbot, QinetiQ Talon robots, invested in the Inuktun series of microrobots, and inspired the Dragonrunner and throwable robots. He wanted to use CRASAR to push for the use of small military robots in emergency response, which was eerily prescient. At the World Trade Center, the 2 years of fieldwork allowed my team to contribute to the response: Our small robots were the most heavily used, we set up the data collection and archiving processes, we provided tools, and we could speak the lingo.

In 2002, I became the director of CRASAR when John resigned to return to military service in Afghanistan. Under my direction, CRASAR has continued to lead the way in deploying robots, including the first deployment of unmanned surface vehicles, which was at Hurricane Wilma in 2004, and the first deployment of small unmanned vehicles at Hurricane Katrina in 2005. The CRASAR cache of tactical ground, aerial, and marine robots remains on call 24/7, and upon my move to Texas A&M University, we established the *Roboticists Without Borders* program to match robots from industry and agencies with disasters. CRASAR also trains emergency managers and responders all over the world about robots and how they can, and have, been used. Other agencies have adopted robotics, and the deployments have begun to accelerate.

Within the academic community, I have been active in establishing disaster robotics and human–robot interaction as a recognized discipline within robotics and automation, cofounding an annual scientific conference, the IEEE International Symposium on Safety, Security, and Rescue Robotics, and hosting numerous workshops and tutorials. I have continued to publish and conduct basic research in my field. These activities have exposed me to other perspectives on rescue robotics and allowed me to witness the growing interest in, and frequent questions about, disaster robotics.

Now, with more than a decade of deployments and research since the collapse of the Twin Towers of the World Trade Center, disaster robotics has matured, and the range of deployments is sufficiently broad to merit a comprehensive examination of the field. As of April 2013, I personally have the most experience with those deployments of robots (15 of 34) and in formally analyzing the performance of robots at those 15 deployments and at 8 other disasters. This gives me a unique position from which to synthesize the state of research and the state of the practice in disaster robotics.

Content and Convention

This book uses several features designed to promote learning and to stimulate further reading and discussion. Each chapter begins with a statement of objectives and ends with a summary to help the reader to find information of value. Chapters 2–5 include a section on common misperceptions about the topic for that chapter, the gaps between theory and practice, and what research directions are suggested by those gaps. Case studies are interspersed, and they illustrate particular technologies and the role of the robots in the larger emergency enterprise.

The book is divided into six chapters, as follows:

• Chapter 1 provides a broad overview of rescue robotics in the context of the larger emergency informatics enterprise, the specific missions, and the major applications of robotics by the type of disaster.
• Chapter 2 presents a summary of the 34 deployments of robots to disasters worldwide in chronological order and offers a formal analysis of how well they have performed their missions and where and why they have failed and the general trends.
• Chapters 3, 4, and 5 describe the typical disaster robot modalities: ground, aerial, and marine, respectively. Each of these chapters follows the same organization of content to enable the reader to compare and contrast the modalities. The chapters repeat and expand on the material from chapters 1 and 2, so that a reader interested only in one modality, say unmanned aerial vehicles, can go directly to the relevant chapter to find

  ▪ the types of robots within that modality;
  ▪ the environments they are expected to operate in, the missions and tasks, and where they have been used;
  ▪ the selection heuristics for determining if that modality is appropriate for a disaster; and
  ▪ the surprises, gaps, and open research questions.
• Chapter 6 describes the types of fieldwork and provides practical advice in planning and conducting fieldwork, what data to collect and how, and working with emergency professionals.

In addition, a list of acronyms and abbreviations and a glossary are found at the back of the book.

The book is written largely in third person but uses a first person point of view in two different situations. First person is used for the case studies in keeping with ethnographic reporting norms and because based on more than 20 keynote addresses and tutorials, I have found that first person storytelling encourages the reader to relate to the realities of disasters and think more closely about how to apply the principles that are distilled in this book. First person is also used to distinguish when material is from the corpus of studies that I as a scientist have contributed to or when the analysis is based on my personal experience. While this book is intended to be a distillation of the entire field and should neutrally reflect the state of the art, my biases most certainly appear, and first person alerts the reader to that possibility.

The book uses boldface and italic to help highlight material. Boldface is used for definitions and important concepts that an instructor may wish to emphasize or test over, and italic is used for emphasis.

Acknowledgments

I have many people and organizations to thank for their help: funding agencies, the two universities I have worked at during the evolution of disaster robotics, the professional responders, team members, my students, the MIT Press, and above all, my family.

This book could not be possible without the research funded by the National Science Foundation (NSF), the Mine Safety and Health Administration (MSHA), the Defense Advanced Research Projects Agency, the Office of Naval Research (ONR), the Army Research Laboratory (ARL), Microsoft Research, the Global Center for Disaster Management and Humanitarian Assistance at the University of South Florida, and the Florida High Tech Corridor. Rita Rodriguez at NSF and Jeff Kravitz at MSHA have been especially supportive. The gifts from SAIC through Dr. Clint Kelly combined with my endowment from Raytheon and donations from SeaBotix to our Roboticists Without Borders program have helped cover the costs of deployments, exercises, and training for responders.

The University of South Florida and Texas A&M University have been wonderfully supportive. My colleagues at the University of South Florida covered my classes for fall 2001 to allow me to analyze the lessons learned from the World Trade Center and travel to promote rescue robotics. Everyone there encouraged me to "go for it," giving me the freedom to help create a new scientific discipline. The faculty and students at Texas A&M remain an inspiration with their ideas, encouragement, and accommodation of my sudden disappearances. I am grateful to work with the Texas A&M Engineering Extension Service, learning from world leaders in all facets of public safety and using its excellent facilities such as Disaster City®.

Countless emergency response professionals have shared their insights and time, but I particularly want to thank the following individuals and groups: Ron Rogers, Hillsborough County Fire Rescue, and Florida Task Force 3 gave generously of their time and expertise, and Mike Gonzalez opened the doors to the Tampa Fire Training Academy and their experts. Bill Bracken actively involved us in structural forensics. Justin Reuter and Indiana Task Force 1 adopted CRASAR after September 11, 2001, and their technical search team manager, Sam Stover, remains an integral part of CRASAR and has organized several of our deployments. No greater love has an Indianapolis urban search and rescue team than to give up going to the Brickyard 400 in order to host a comprehensive search and rescue exercise for 20 scientists for our first Summer Institute. Jim Bastan with New Jersey Task Force 1 became the first adopter of rescue robots in the United States and remains a source of great ideas. New Jersey's disaster training facilities remain my second favorite place on Earth, just second to Disaster City®. John Holgerson and Joe Sorrentino of Rescue Training Associates incorporated us into training events at major demolitions all over the country, often with a large cadre from Florida Task Forces 1 and 2 out of Miami; it is now difficult to conduct a response exercise without a shot of Cuban coffee. Geoff Williams from the United Kingdom, who remains an inspiration, and Jacques du Plessis at Rescue South Africa engaged us on an international scale. California Task Force 2 and California Task Force 1 hosted us for a rescue robotics awareness training exercise for the Los Angeles basin, followed by Los Angeles County Fire Rescue training us on their confined-space rescue techniques. The U.S. Marine Corps Chemical Biological Incident Response Force (CBIRF) continues to educate me on hazmat responses, and the 10 days I spent for their CBIRF boot camp has had a major impact on how I conceptualize the diversity of roles of robots for disaster response. Virginia Task Force 2 and Missouri Task Force 1 also have been great. The response experts at the Texas A&M Engineering Extension Service and Texas Task Force 1 have also been helpful, with Billy Parker offering insights into where and how robots could be deployed long before I moved to Texas A&M. Kem Bennett, who was the dean of engineering at Texas A&M, the founder of Texas Task Force 1, and the creator of Disaster City®, remains a role model. Clint Arnett is our go-to guy and has organized countless exercises, Summer Institutes, and experiments at Disaster City® and has deployed to Cologne, Germany, with me—how it all works is a mystery but a fun one. Bob McKee co-hosted several Summer Institutes at Disaster City® with me and served as a co-principal investigator on an NSF grant, and David Martin, Susan Brown, and Matt Minson have been highly supportive.

This book also benefits from the experiences of everyone who has participated in deployments, but Sam Stover and Eric Steimle deserve special thanks. They help me organize resources and responses, and they volunteer their time and talents in keeping up with the best systems that are emerging from industry and in promoting rescue robotics. I would like to thank all the people I've deployed with: Fred Alibozek, Mike Bruch, John Blitch, Jennifer Casper, Mike Ciholas, Bart Everett, Tom Frost, Jay Haglund, Robin Laird, Arnie Mangolds, Mark Micire, Brian Minten, Grinnell Moore,

Gary Mouru, Chris Norman, Scott Pratt, Andy Shein, Chris Smith, (World Trade Center), Sam Stover (La Conchita mudslides), Mark Micire, Sam Stover (Hurricane Charley), John Dugan, Chandler Griffin, Mike Lotre (Hurricane Katrina), Chandler Griffin, Kevin Pratt, Sam Stover (post–Hurricane Katrina), Charlie Cullins, Chandler Griffin, Mark Hall, Kevin Pratt, Eric Steimle (Hurricane Wilma), Ron Gust, Jennifer McMullen, Sam Stover (Midas Gold Mine collapse), Jeff Kravitz, James Milward, Ken K. Peligren, Jeff Stanway (Crandall Canyon Mine collapse), Bill Bracken, Jenny Burke, Jeff Craighead, Chandler Griiffin, Noah Kuntz, Kevin Pratt, Toshihiko Nishimura, Sam Stover, Osachika Tanimoto, (Berkman Plaza II collapse), Stefan Hurlebaus, Michael Lindemuth, Daniel Mays, Zenon Medina-Cetina, Eric Steimle, David Trejo (Hurricane Ike), Clint Arnett, Sebastian Blumenthal, Thomas Christaller, Thorsten Linder, Peter Molitor, Hartmut Surmann, Satoshi Tadokoro (Cologne Historical Archives collapse), Daniel Nardi (L'Aquila earthquake), Karen L. Dreger, Michinori Hatayama, Osamu Kawase, Tetsuya Kimura, Kazuyuki Kon, Kenichi Makabe, Fumitoshi Matsuno, Hisashi Mizumoto, Sean Newsome, Jesse Rodocker, Brian Slaughter, Richard Smith, Satoshi Tadokoro (Tohoku tsunami), and Lindsey Ballard, Scott Bump, Jeff Lumpkin, and Bradley Welch (Fukushima Daiichi).

I want to thank my students and postdoctoral students for continuing to teach me as much as I teach them and for the great times we have exploring new ideas and insights. John Blitch, Jenny Burke, Jennifer Casper, Jeff Craighead, Mark Micire, Brian Minten, and Kevin Pratt have been extraordinary in the gifts of their time and hearts to the humanitarian use of robots and have deployed to disasters. Cindy Bethel, Brian Day, Brittany Duncan, Jennifer Carlson, Thomas Fincannon, Rachel Flores-Meath, Aaron Gage, Marco Garzo, Zack Henkel, Jeff Hyams, Elliot Kim, Mike Lindemuth, Matt Long, Josh Peschel, Tanner Perkins, Brandon Shrewsbury, Vasant Srinivasan, Jesus Suarez, and Rosemarie Yagoda have spent countless hours at exercises and in the field and helping to support teams at deployments. Marco Garza, Edgardo Calzado, and Tanner Perkins have served as student laboratory managers with the hidden but essential task of keeping CRASAR ready to go at a 4-hour notice.

Marie Lee, Ada Brunstein, and Bob Prior at the MIT Press have guided the publishing of this book and the anonymous reviewers provided excellent feedback. Christopher Curioli, who copy edited this volume for the MIT Press, was outstanding; he helped the book maintain a consistent pedagogical focus and counteracted my tendency to overuse italics for emphasis which had created a reading experience suitable for delivery by William

Shatner, where everything (pause) is (pause) important. I especially want to thank Brittany Duncan, Rachel Flores-Meath, Zachary Henkel, Elliot Kim, Tanner Perkins, Vasant Srinivasan, Brandon Shrewsbury, and Jesus Suarez for their critiques of early drafts and numerous contributions to the overall organization and figures.

I could not have written this book or conducted the work upon which it is based without the support of my family. After the deployment to the September 11 disaster, it was clear that there needed to be an active advocate of rescue robotics for adoption and for researchers and industry to recognize the special demands of emergency response. The other CRASAR participants were being re-absorbed into their day jobs, so either I was going to have to step up and be the advocate or no one would. However, that would entail a huge cost in time and travel; indeed I gave more than 50 talks on rescue robotics to agencies and at conferences in the first year alone. My husband, Kevin, our two children, Kate and Allan, and I made the decision together, and they have been totally supportive during my long hours away from them. I could not do any of this without Kevin.

Any errors herein are solely mine, and I fervently hope they are minor and do not distract or mislead anyone in the quest to make and use robots to assist in disaster prevention, preparedness, response, and recovery.

1   Introduction

Rescue robotics is devoted to enabling responders and other stakeholders *to sense and act at a distance from the site of a disaster or extreme incident*. Although the high public profile of search and rescue robots counters the weaponized-drone portrayals of robots, rescue robotics is a relatively small discipline within corporate and academic circles. However, as will be discussed later, the impact of earthquakes, hurricanes, flooding, and accidents is increasing, so the need for robots for all phases of a disaster, from **prevention** and **preparation** to **response** and **recovery**, will increase as well. This broader consideration of robots for all phases of a disaster and for disaster science in general is referred to as **disaster robotics**, though the terms *search and rescue robot*, *rescue robot*, and *disaster robot* will be used interchangeably. As robots evolve to handle extreme events, they will hopefully find their way from the military to local fire rescue departments and other stakeholders for use in "regular" emergencies.

This chapter addresses five basic questions to provide background for the material that will be covered by this book:

- *What is a rescue robot?* This is a broad question with a set of related questions to be addressed: What makes rescue robots different from "regular" mobile robots? Aren't military robots sufficient? How did rescue robotics get started?
- *Why rescue robotics?* While robots (or any mechanism to save lives and reduce suffering) have an emotional appeal, is there really an economic justification for rescue robots or are incidents too rare to justify the expense? Can't animals or insects be used; after all, they are much more mobile than any robot?
- *Who uses rescue robots?* This is a question about the nature of who is involved in disasters: Are robots used and owned by federal or state

authorities? Local authorities? Volunteer organizations? Can or do companies and private citizens use robots?
- *When can rescue robots be used?* This is another question about the nature of responses: Are robots just for use after a disaster or are they (or can they be) used before?
- *What are rescue robots used for?* The popular expectation is that robots are used for immediate lifesaving operations such as searching for survivors, but what are the other critical tasks that robots can help with?

This chapter begins with a brief history of the origins and development of the field, then proceeds to describe what a rescue robot is compared with other kinds of robots. Rescue robots are tactical, organic, unmanned systems that allow emergency professionals to perceive and act at a distance in real time, also known as having a remote presence at the disaster site. The chapter then describes why robots—and the remote presence they provide—are needed to assist in saving lives, reducing long-term health consequences, and accelerating economic and environmental recovery. Surprisingly few formal response organizations own rescue robots, which explains the average lag time of 6.5 days for a robot to be used at a disaster. The chapter then turns to an emergency management perspective on robotics, describing the phases of a disaster and then the types of disasters and associated missions.

1.1   Brief History

The general consensus is that the use of robots from the Defense Advanced Research Projects Agency (DARPA) Tactical Mobile Robots program by the newly established Center for Robot-Assisted Search and Rescue (CRASAR) at the World Trade Center disaster in New York in 2001 represents the first known use of rescue robots. Yet, speculation as to the utility of robots for search and rescue, firefighting, and other emergency missions began in the 1980s (Kobayashi and Nakamura, 1983). Academic research into rescue robotics started in 1995 (Davids, 2002), most notably led by two research groups. Prof. Satoshi Takokoro's group, then at Kobe University in Japan, was motivated by the tragic loss of life in the Hanshin-Awajii earthquake in Kobe, Japan, in 1995; his group became the foundation for the International Rescue System Institute (IRS). My laboratory, then at the Colorado School of Mines in the United States, was motivated by the 1995 bombing of the Alfred P. Murrah Federal Building in Oklahoma City, Oklahoma. Two mobile robot competitions, the Association for the Advancement of Artificial Intelligence (AAAI) Mobile Robot Competition in the United States and the RoboCup Rescue League internationally, were started shortly thereafter to engage the scientific community in rescue research. The RoboCup Rescue League continues, with its focus shifting from enabling research to providing students with a socially relevant set of tasks within which to build robots and explore artificial intelligence principles.

The site of the 2001 collapse of the World Trade Center buildings, however, was not the first site where robots were used at a disaster. Dr. Red Whitaker at Carnegie Mellon University constructed robots to enter and explore the Three Mile Island nuclear facility several months after the 1979 incident and also constructed robots for the 1986 Chernobyl disaster. These robots were thickly armored to protect against radiation, which made them heavy, large, and fairly slow. Fortunately, both nuclear disaster areas were **human scale** (i.e., a site large enough and still intact enough that a human outfitted in protective gear may walk through the site), and thus robot size was not an issue. Likewise, robot speed was not critical, as the robots were being used for postdisaster mitigation and not for time-critical lifesaving or injury stabilization. The robots required multiple operators for teleoperation. The success of these robots set the precedent for nuclear robots as radiation-hardened, large, heavy, and teleoperated systems. These designs began to migrate to the bomb squad community, where the heavy shielding protected the expensive robot from explosions, and speed was not a factor.

The tipping point in the evolution of current rescue robots came at the Oklahoma City bombing in 1995, where numerous bomb squad robots sat idle (Manzi, Powers, and Zetterlund, 2002), too big to navigate the narrow voids and too heavy to use without fear of causing a secondary collapse that might kill trapped survivors or rescue workers working in layers below the surface. The large, ponderous robots created for nuclear disasters and bomb squads eventually generated a backlash in the mobile robot community, which led to newer ground robots, including rescue robots, being conceived as small, agile, and autonomous with a small degree of supervision by a single human. The advances in robotics and artificial intelligence in the 1990s—especially the smaller platforms being developed for the NASA Mars missions and reactive behavioral software architectures—supported the moral imperative for roboticists to contribute to this societal need.

Advances in usable rescue robots have been partially enabled by investment by the U.S. military, especially through DARPA. The DARPA Tactical Mobile Robots (TMR) program under the program management of Lt. Col. John Blitch was highly influential in creating smaller robots that could be carried in backpacks and operated by one or two soldiers. The iRobot Packbot and QinetiQ mini-Talon are perhaps the best-known outcomes of this program. While the intended application was military operations in urban terrain, the potential for dual use of TMR robots with rescue operations was also a consideration (Krotkov and Blitch, 1999; Blitch, Murphy, and Durkin, 2002), leading Lieutenant Colonel Blitch to found CRASAR shortly before the September 11 attack on the World Trade Center. Developments in small unmanned aerial systems for Special Operations forces led to their use in the aftermath of Hurricane Katrina, and the National Guards of several states on the U.S. Gulf Coast use the larger Predator unmanned aerial vehicles for surveying large areas before and after a hurricane. With the use of small, agile (but not radiologically hardened) ground robots such as the iRobot Packbot and QinetiQ Talon and aerial vehicles such as the Honeywell T-Hawk during the Fukushima Daiichi nuclear emergency, rescue robots have returned to their nuclear origins.

However, commercially available ground and marine robots designed for civilian uses also have a dual use for disasters. One frequently used type of ground robot at disasters is the pipeline inspection robot, such as the Inuktun Micro-Tracks and Micro-VGTV series; these robots are typically much smaller than military systems and can penetrate into voids that people, dogs, and robots designed to dispose of improvised explosive devices cannot enter. The other frequently used ground robot is the U.S. Mine Safety and Health Administration's mine-permissible variant of a Remotec civilian bomb squad robot. Unmanned marine vehicle technology stems primarily from oceanographic research and the underwater inspection business.

1.2   What Is a Rescue Robot?

Rescue robots are **mobile robots**, or robots built to sense and act in an environment; in artificial intelligence terms, this type of robot is often called a **physically situated agent**. Mobile robots are distinct from factory robots and industrial manipulators that perform repetitive tasks. By moving about in the often unpredictable world, these robots have to be more aware of the environment, not just for navigation but also to make sure that they aren't causing secondary collapses, dislodging debris, or disturbing forensic evidence.

Rescue robots are a category of mobile robots that are generally small enough and portable enough to be transported, used, and operated on demand by the group needing the information; such a robot is called a tactical, organic system, compared to a strategic asset such as a Predator or a Global Hawk. A rescue robot may be designed differently than a military unmanned system because of the need to meet three design constraints:

1. *extreme terrains and operating conditions* that affect size, sensor performance, and pose general robot survivability constraints;
2. *ability to function in GPS- and wireless-denied environments*; and
3. provision of *appropriate human–robot interaction* for operators "behind the robot" and for victims "in front."

The history of rescue robots illustrates some of the interplay between robots developed for defense and space applications and rescue robotics.

1.2.1   It Is a Tactical, Organic, Unmanned System

Mobile robots are often referred to as **unmanned systems** to distinguish them from robots used for factory automation. Unmanned systems have three different **modalities** based on where they operate. If they operate

- *on the ground*, they are called **unmanned ground vehicles** (UGVs);
- *in the water*, they fall in the general category of **unmanned marine vehicles** (UMVs);
- *in the air*, they may be called by various names including **unmanned aerial vehicles** (UAVs), **unmanned aerial systems** (UASs), **remotely piloted aircraft** (RPA), or **remotely piloted vehicles** (RPVs).

The three different modalities of robots have been developed by independent communities, leading to somewhat contradictory marine and aerial terminologies and taxonomies. UMVs may operate on the surface of water, underwater, or, more rarely, in both environments. UMVs restricted to the water surface are often called **unmanned surface vehicles** (USVs). UMVs working underwater may be referred to as **unmanned underwater vehicles** (UUVs). UUVs fall into two main categories: tethered **remotely operated vehicles** (ROVs) and untethered, free-swimming vehicles called **autonomous underwater vehicles** (AUVs). The letters in AUV are sometimes rearranged to UAV for "underwater autonomous vehicle," leading to confusion with "unmanned aerial vehicle." Unmanned aerial vehicles are sometimes called UASs (to emphasize the sociotechnical system aspects) or RPVs (to emphasize that there still is a human in the loop). Regardless of nomenclature, UAVs fall into two categories: fixed wing (e.g., looks and acts like a plane or jet) and vertical takeoff and landing (VTOL; e.g., looks and acts like a helicopter).

Regardless of robot modality, this book restricts rescue robots to unmanned systems used as *tactical, organic assets* (figure 1.1). **Tactical** means the robot is directly controlled by stakeholders with "boots on the ground"—people who need to make fairly rapid decisions about the event. **Organic** means that the robot is deployed, maintained, transported, and tasked and directed by the stakeholder, though, of course, the information can be shared with other stakeholders, and via networks the stakeholders may be far removed from the site of operation. A **tactical, organic asset** travels with the stakeholder and thus is immediately available and used on demand (and also tends to be smaller and more portable). This is in contrast to a **strategic asset**, such as a satellite, a military or Coast Guard helicopter, or a high-altitude drone, which is deployed and tasked by an agency that then passes or filters information to the tactical teams as it becomes available. There are exceptions to every rule, and it is possible to have a strategic unmanned system used tactically: NASA's Ikhana high-altitude Global Hawk was directed in real time by the California wildfire incident commander on the ground at the 2006 California wildfires (Ambrosia et al., 2010). But, in general, Ikhana is not considered a rescue robot.

1.2.2 It Satisfies Three Design Constraints

Disaster robotics is similar to robotics for military applications and historically has leveraged defense and civilian inspection robots, but it is the need to satisfy three constraints simultaneously that makes disaster robots different from robots designed for other applications. Rescue robots must be small enough and environmentally hardened to function in extreme terrains and operating conditions, able to work in environments where GPS and wireless signals are blocked, and must interact with victims and other responders "in front" of the robot as well as with the operator and stakeholders "behind" the robot.

Disasters present **extreme terrains and operating conditions** that affect *size, sensor performance, and general robot survivability* (figure 1.2). As will be described in later chapters, UGVs must function in openings as small as 3 centimeters (e.g., Berkman Plaza II collapse) and must move horizontally and vertically through irregular voids and uneven mixtures of corrosive building materials and textured furnishings. UGVs are often exposed to large amounts of groundwater (e.g., Crandall Canyon Mine and Pike River Mine). The robots may have to navigate through mud or drilling foam, which fouls sensors (e.g., Crandall Canyon Mine) and interferes with effectors (e.g., La Conchita mudslides). UGVs may have to operate in extreme heat (e.g., World Trade Center) or explosive atmospheres (e.g., Sago Mine and Pike River Mine). UMVs may have to function in currents, avoid flotsam and debris, avoid complications from tides and flood crests, and peer through turbid waters. UAVs may have to overcome unpredictable buffeting and wind shears near urban structures. Any of these types of vehicles may need to function in smoke (e.g., World Trade Center) or be shielded to function while exposed to radiation (e.g., Fukushima Daiichi).

Figure 1.1
Typical rescue robots are small and easily portable: (top left) Inuktun VGTV Xtreme
ground robot (Berkman Plaza II collapse); (top right) iSENSYS miniature helicopter
(Hurricane Katrina); (bottom) YSI Echomapper and SeaBotix SARbot marine vehicles
(Tohoku tsunami).

Rescue robots typically function in **GPS- and wireless-denied environments**. The material density of commercial buildings interferes with GPS
and wireless networks for robots working in the interiors of commercial
buildings. However, interference is not limited to interiors; urban structures
such as buildings or bridges can create shadows or multiple paths that affect
approaching UAVs and UMVs. Inertial navigation sensors are often large,

Figure 1.2
Example of the extreme terrain that a robot is expected to enter and travel *through*, not over (Historical Archive of Cologne building collapse).

expensive, and susceptible to sudden movements (such as the bump as a robot slides over a rock). Wireless communications can be boosted with more power, but that leads to a trade-off between making the robot larger to carry the power and it becoming too large to be of use.

Rescue robots also present **extreme human–robot interaction** challenges for operators and for victims. Stakeholders in the safe **warm** or **cold zones** use rescue robots to help them see and act in the **hot zone**, the restricted area of impact. This means the robots mediate between the operator and the distal environment. Mediation introduces cognitive challenges, such as trying to see and understand the world through the robot's "keyhole" view into the situation; this can lead to poor performance and errors. Teleoperation assumes the operator can overcome these cognitive challenges, though usually with the help of well-designed user interfaces. Artificial intelligence has focused on making the robot completely autonomous, though a middle ground of adding intelligence assistance seems more productive. NASA pioneered the field of teleoperation, which investigates remote presence primarily in terms of mitigation of time delays in transmission of data over long distances. In rescue robotics, the conditions for remote presence are more demanding than those for NASA space operations because missions do not have days or months of preplanning, the world may dynamically change due to a secondary collapse or a levee break, and the amount of training and rehearsal is much less. However, human–robot interaction challenges are not limited to operators interfacing with the robots; victims will also interact with rescue robots. A comprehensive study demonstrated that if robots operate without considering a victim's personal space, the simulated victim's biophysical measurements show increased stress and discomfort (Bethel and Murphy, 2010).

1.3   Why Rescue Robotics?

Disaster robots help stakeholders prevent, prepare for, respond to, and recover from the increasing number and complexity of urban disasters and extreme events by providing stakeholders with the ability to access an area of interest. If a disaster has occurred, it may be physically impossible, too dangerous, or too inefficient for a responder to enter the hot zone. Robots do not replace people or dogs but rather complement human and canine abilities and mitigate their risk. Disasters have a significant impact on lives, quality of life, the economy, and the environment, and if they cannot be prevented, rapid response means less loss of life and long-term injuries and faster economic and environmental recovery.

1.3.1   The Impact of Urban Disasters and Extreme Incidents Is Increasing

The U.S. Federal Emergency Management Agency (FEMA) *Guide to Emergency Management and Related Terms, Definitions, Concepts, Acronyms, Organizations, Programs, Guidance, Executive Orders & Legislation* gives multiple definitions of the term *disaster*; this book will use "An event that requires resources beyond the capability of a community and requires a multiple agency response" as the definition of **disaster** (Blanchard, 2007). Unlike a disaster, an **extreme incident** is handled internally within a fire rescue department, usually by a team with specialized training. As a result, extraction of workers trapped in a cave-in at a construction site may be addressed by the urban search and rescue (US&R) specialists in the local fire rescue department, while the collapse of a major building complex such as the World Trade Center in 2001 may require the aggregation of US&R specialists from all over the United States, as well as additional expertise.

Of all the types of disasters, events in urban settings are the most significant in terms of deaths and economic impact. The most recent *World Disasters Report* (McClean, 2010) prepared by the International Federation of Red Cross and Red Crescent Societies reports that between 2000 and 2009, 1,105,353 people were killed worldwide in 7,184 disasters with another 2,550,272 more directly affected, creating a cost of $986.7 billion. The number of disasters each year remained relatively constant, around

700. However, the impact of the disasters was a function of whether or not the event occurred in an urban area. The *World Disasters Report* argues that deaths and effects of disasters will dramatically increase, citing the increasing numbers of people living in concentrated urban environments, as well as the consequences of damage to urban structures where people live and work (e.g., office and apartment buildings) and to structures serving to mitigate a disaster (e.g., hospitals, transportation, and other critical infrastructure).

The statistics from the 1985 Mexico City earthquake and subsequent urban disasters (United States Fire Administration, 1996; National Fire Protection Association, 1999) established the conventional wisdom that *only a small fraction of trapped victims actually survive an urban incident*. Eighty percent of survivors of urban disasters are **surface victims**; that is, the people lying on the surface of the rubble or readily visible to a rescuer. Only 20% of survivors of urban disasters come from the interior of the rubble, yet the interior is often where the majority of victims are located. Those victims may be in (relatively) good health and **entombed** or **trapped** beneath rubble with urgent medical needs. Worse yet, the mortality rate increases and peaks after 48 hours, meaning that survivors who are not extricated in the first 48 hours after the event are unlikely to survive beyond a few weeks in the hospital. Others will be crippled or have persistent health issues. These statistics suggest that robots that can penetrate into the interior of rubble to speed search, rescue, and intervention would have a significant impact.

But the value of robots for economic and environmental recovery should not be ignored. For example, our use of ROVs at Minamisanriku in Japan 6 months after the Tohoku tsunami to assist in clearing Shizugawa Bay of debris had no lifesaving value. But consider that Minamisanriku is the second largest fishing area in Miyagi, a prefecture famous for its fishing industry and pure waters. The Shizugawa Fishing Cooperative was eager to remove submerged cars and the more than 800 fishing boats (about 85% of the fleet) known to be missing (Leitsinger, 2011) that might be leaking fuel and contaminating the water; to remove rubble that might snag nets; and to reopen navigation beyond a few major channels. The loss of fishing affected the availability and price of food for everyone in Japan, but it also affected the local community: fishermen who routinely made $86,000 to $124,000 a year were reduced to making $100 to $150 a day collecting wreckage (Leitsinger, 2011) until fishing was restored.

1.3.2 Robots Provide a Remote Presence at the Disaster

Stakeholders need the ability to "sense and act at a distance," which can be more formally stated as they want to have a **remote presence** in the incident area (Murphy and Burke, 2008). Each disaster is different and cannot be accurately modeled. Thus, a robot can enable a stakeholder to see, in real time, deep into the interior of a collapsed building where there is no physical entry large enough for a human or dog or where the area is unsafe or does not support life (e.g., on fire, engulfed in hazardous materials or radiation). Robots can penetrate further than the 3 to 4 meters a camera on a wand can extend into densely packed debris. They can interact with the environment to mitigate a disaster; for instance, the use of robots to insert a containment cap at the 2010 Deepwater Horizon explosion (Newman, 2010) or to turn valves at the 2011 Fukushima Daiichi nuclear incident (Strickland, 2011).

1.3.3   Rescue Robots Work Where People and Animals Cannot

Rescue robots serve as smart tools that complement, not replace, people and animals. Robots can stay on task indefinitely (depending on power source) and do not tire. They can go into places and do combinations of things or host sensors that animals cannot. Robots can be turned off and stored indefinitely, unlike animals. Robots do not require special handling in the way that a swarm of wasps for smelling bodies (Science Daily, 2006) might, as they do not require food to be brought with them or create bio-waste to be disposed of. Most importantly, they are not alive. Immediately after the World Trade Center disaster, rats with camera implants were posed as an alternative to expensive microrobots for navigating through densely packed rubble. But as noted in Murphy (2002), these "robo-rats" could not have survived the high temperatures from the smoldering fires in the interior of the World Trade Center rubble that the robots worked in. (Not to mention terrorizing trapped survivors, which was the subject of a segment of the *Daily Show* with Jon Stewart.)

Likewise in some missions, such as a chemical or nuclear emergency, the robot is expected to do tasks a person could normally do (e.g., turning a valve) and in environments that a person could normally work in. But the robot is not replacing a responder, because the responder could not have lived long enough, even with protective gear, to accomplish the task or would have died shortly afterward from the effects. *To be clear, responders expect to continue to risk their lives to save others; robots don't replace their rapid actions or sacrifices but rather eliminate unnecessary risks.*

1.4   Who Uses Rescue Robots?

Rescue robots are or can be used by a wide variety of agencies, nongovernmental organizations (NGOs), and even industries and companies. The definition of rescue robotics points to "responders and other stakeholders" as the end users of rescue robotics, not robot operators. Emergency management is best described as a **polycentric** enterprise, where there are multiple ("poly") places or organizations ("centers") that integrate or share information and that can bring expertise or capability to the situation (Andersson and Ostrom, 2008). Regardless of the incident command hierarchy specified in documents such as the U.S. National Response Framework, emergencies always include some *surprising and unique demands*, because multiple organizations become involved in a response *each with a different mix of response expertise and capabilities* and because the time course of response and recovery brings a *changing mix of different organizations* together. To see one example of how even a straightforward incident is polycentric, consider the 2010 Deepwater Horizon explosion and oil leak. While BP retained authority for the disaster and the U.S. Coast Guard provided important expertise, there was nothing in prior planning that factored in the role of local fishing and tourism industry organizations as stakeholders. Robots are a means, not an end, to enabling each group to get the data they need to make effective decisions.

Only two agencies in the United States own rescue robots. New Jersey Task Force 1, a state team, not a FEMA team, is the only US&R team in the United States known directly to own and maintain a rescue robot. Federal US&R teams are not allowed to purchase rescue robots because they have not been approved. The U.S. Department of Homeland Security, which FEMA is part of, has been sponsoring the creation of standard test methods that would verify that a robot model meets the minimum qualifications for a rescue robot. No Japanese agency owns a robot, though British fire rescue departments have been actively acquiring all modalities of rescue robots. The U.S. Mine Safety and Health Administration (MSHA) owns the only robot in the world certified to work in explosive atmospheres.

While robots have become ubiquitous in U.S. military operations, they are still quite new to disaster response. This explains in part the average lag time of 6.5 days between a disaster and when a robot of any modality is used: If the incident command institution had a robot or an existing partnership with a group that had robots, there was an average lag time of 0.5 day before the robot was used, whereas if not, the average lag time went up to 7.5 days. The socio-organizational culture of response and adoption constraints will be discussed in chapter 6.

1.5   When Can Rescue Robots Be Used?

The short answer to "when can rescue robots be used?" is anytime, though researchers have focused on development of systems primarily for the immediate response phase rather than for before the event occurs or for mitigation and recovery. A related question is *when have they been used?* The short answer to that is late in the response or recovery. Robots are deployed on average 6.5 days after the event, reducing their immediate value for lifesaving and reconnaissance, but they are still useful for other missions. Chapter 6 will discuss more about the socio-organizational aspects of a disaster that lead to these delays.

Since the 1970s, the **disaster life cycle** has been thought of as having **four phases**: *prevention, preparation, response*, and *recovery* (Blanchard, 2007). Each phase conjures up expectations of different agencies; for example, the U.S. Forest Service created Smokey the Bear to encourage prevention of forest fires; response is associated with urban search and rescue and National Guard teams; and recovery is associated with insurance inspectors and business recovery loans. In reality, these entities are involved throughout the entire disaster life cycle.

These phases are not independent and sequential so that several stakeholders may each have robots or the data from a robot deployed by one agency may be of immediate use to another agency. Response and recovery operations start instantaneously, though the more visible and emotionally compelling lifesaving mission of rescuers often dominates media reports. Response operations rarely last beyond 10 days, though local officials may be reluctant to declare the response phase over. Initial recovery operations can go on for months and are generally marked by when roads, schools, and hospitals are reopened. Economic recovery often takes years beyond that.

With response and recovery occurring simultaneously, a single robot can provide information to many "consumers." An unpublished study by the author and colleagues at Ohio State University on the role of robots for structural inspection identified that urban structures are likely to be manually inspected at least four times by different stakeholders from the response and recovery groups: search and rescue teams to determine safe entry for search, the American Red Cross to estimate how long a region will need assistance, insurance adjusters, and building inspectors.

Rescue robots are typically deployed fairly late, toward the middle of the response phase or later. My analysis of robot deployments worldwide in 2010 (Murphy, 2011b) showed that the average time between an incident and the actual use of a robot was 6.5 days. If the analysis considers only the five deployments where the mission was clearly to search for survivors (e.g., Upper Big Branch Mine; Wangjialing Coal Mine; Haiti earthquake; Prospect Towers; Pike River Mine), then the average was 4.2 days for a robot to arrive, well after the 48-hour peak in the mortality curve—too late to be of value. The biggest predictor of whether a robot would be deployed and how quickly was whether the agency in charge had a robot or a partner with robots. In four of the 2010 deployments, the agency or industry that held incident command responsibility either already used the robots in day-to-day operations (BP at Deepwater Horizon; Italian Coast Guard for a missing balloonist) or prior lines of authority were already in place [MSHA for Upper Big Branch Mine; New Jersey's regional Urban Area Security Initiative (UASI) teams at Prospect Towers]. In those four cases, robots arrived on the scene and were put to use in one-half day on average.

1.6   What Are Rescue Robots Used For?

The previous sections have provided a sense of what robots are and the phases of a disaster, which sets a foundation for understanding the specific missions that rescue robots can be used for. This section first describes the types of disasters followed by 13 proposed missions for rescue robots.

1.6.1   Types of Disasters

There is no comprehensive list of the types of disasters and incidents, perhaps because new events continually emerge. The Federal Emergency Management Agency (FEMA, 2013) offers examples of disasters that are traditionally managed by the fire rescue community, and the list continues to grow. Many are "natural disasters" stemming from **meteorological and geological events** such as *avalanches, earthquakes, floods, fires, heat waves, hurricanes, landslides, thunderstorms, tornadoes, tsunamis, volcanoes, wildfires,* and *winter storms*. Others are **accidents** or **terrorism involving man-made facilities** such as *building, bridge, or tunnel collapses, chemical emergencies, dam failures, hazardous materials, nuclear power plants,* and *train wrecks*. High-profile incidents include *wilderness search and rescue*, where a person(s) may be lost in the woods or trapped in a cave, and *swift-water rescue*. A commonplace urban incident is a *trench collapse*, when construction excavation collapses on workers. In general, management of events involving man-made facilities, with the exception of nuclear power plant emergencies, are collectively referred to as urban search and rescue.

   In the United States, **mining- and mineral-related disasters**, either geological events or accidents, are managed by the owner, not the government. For example, the MSHA does not manage a mine collapse the way that a fire department assumes control of a fire, but the agency is present to provide support. While in theory MSHA or the Coast Guard could assume incident command or be appointed to do so, this is not the case for federal and state agencies. The tradition of assisting rather than assuming command most likely originated from the unique expertise associated with a mining or drilling incident: the companies were more likely to be knowledgeable on how to respond effectively. However, this has led to the notable opacity in recent private mining, drilling, and nuclear responses. There is a potential conflict of interest of a company to minimize response expenditures or to hide potential evidence that could be used against the company in lawsuits, which could retard adoption of rescue robots.

1.6.2 Missions

The entire set of tasks for rescue robotics is incomplete, as the field is new and evolving. It is hard to project all the tasks for a robot, as they do things that humans and animals cannot; instead, imagination and opportunity drive creation of novel uses. This makes most of rescue robotics a **formative work domain** (Vicente, 1999) as opposed to a **normative work domain**, in which robots perform existing tasks (though some missions such as hazardous material response has tasks currently performed by responders that could be done by robots). The *Handbook of Robotics* identifies 10 tasks for rescue robots for the rescue phase (Murphy et al., 2008b), while deployments to recent disasters such as the Tohoku earthquake and tsunami and the Fukushima Daiichi nuclear emergency suggest that robots designed for rescue can support a rich set of tasks for mitigation and recovery as well.

Robots have been used or proposed for the 13 activities in the list that follows, and more activities are expected to emerge as robots are put into use. Training manuals often describe response tasks as being executed serially where one activity follows another (e.g., search then extrication), but emergency workers multitask, performing activities concurrently (e.g., mapping and structural inspection while searching) (Casper and Murphy, 2002). As a result, robots will be expected to be reconfigurable and reused for serial tasks and be able to perform several of the tasks below in parallel.

Search is a concentrated activity in the interior of a structure, in caves or tunnels, or in the wilderness and aims at finding a victim to extricate and any potential hazards. The objective for using robots for the search task is to perform a search where people or existing tools cannot be used, to perform the search faster, and to ensure completeness without increasing risk to victims or rescuers.

Reconnaissance and mapping provides responders with general situation awareness and an understanding of the destroyed environment. The objective is speedy coverage of an area of interest at the appropriate resolution.

In meteorological and geological disasters, the primary focus is on the exterior and outdoors (Where is the damage? Where are people in distress? What is the wildfire doing?), while for hazardous material or nuclear events the focus may shift to what is going on inside of the facility.

Rubble removal, either for extrication of victims or for rebuilding, can be expedited by robotic machinery or exoskeletons. The objective is to move heavier rubble faster than could be done manually, but with a smaller, lighter footprint than required by a traditional construction crane. Robots and exoskeletons may be man-sized or smaller and easier to move to the site and begin working.

Structural inspection and **forensics** are related missions that can be facilitated through robots that deliver structural sensor payloads to more favorable viewing angles. The objective is to enable an engineering understanding of critical infrastructure that affects public services such as bridges, roads, utilities, aqueducts, or shipping channels and urban structures such as buildings or manufacturing inlets and outlets. Structural inspection may be conducted either from the interior (e.g., inserting a robot into rubble to help rescuers understand the nature of the collapse) or on the exterior (e.g., to determine whether a structure is safe to enter).

In situ medical assessment and intervention are needed to permit medical personnel to triage victims and provide life support functions such as transporting and administering fluids and medication. The inability to provide medical intervention was a major problem at the Oklahoma City bombing (Barbera, DeAtley, and Macintyre, 1995). The objective is to provide telepresence for medical personnel during the 4 to 10 hours that it usually takes to extricate a victim (United States Fire Administration, 1996; National Fire Protection Association, 1999).

Medically sensitive extrication and evacuation of casualties may be needed to help provide medical assistance while victims are still in the disaster area, also known as the hot zone. The objective is to ferry survivors rapidly and safely to a collection point. In the case of a chemical, biological, or radiological event, the number of victims is expected to exceed the number that can be carried out by human rescuers in their highly restrictive protective gear; this makes robot carriers attractive. Because medical doctors may not be permitted inside the hot zone, which can extend for kilometers, robot carriers that support telemedicine may be of huge benefit.

Acting as a mobile beacon or repeater in order to extend wireless communication coverage and bandwidth and to localize personnel based on their radio signal strength. The objective is to ensure reliable communications for both robots and the response and recovery operations in general.

Serving as a surrogate for a team member such as a safety officer or a logistics person, where team members outside the hot zone use telepresence to work with team members in the hot zone to monitor the state of progress and anticipate needs. The objective is to use robots to speed up the workflow and to help notice developing events that could pose a threat to workers concentrating on their immediate tasks.

Adaptive shoring of unstable rubble to expedite the extrication process. Rubble removal is often hindered by the need to take a conservative pace in order to prevent a secondary collapse that might further injure a trapped survivor or responder. The objective is to maintain the state of the collapse without being able to map out the structure fully.

Providing logistics support by automating the transportation of equipment and supplies from storage areas to teams or distribution points within the hot zone. The objective is to reduce the number of personnel in the hot zone devoted to routine tasks.

Victim recovery support is needed after the possibility of survivors remaining has dwindled and operations turn to extricating bodies already found or finding all the missing. The objective is to search exhaustively and help extricate all remaining victims.

Estimation of debris volume and types to speed cleanup and enable residents to reenter affected areas. Meteorological and geological events distribute trees, homes, cars, and other materials throughout the hot zone. This not only blocks roads but also can cause environmental and sanitation problems, preventing residents from returning until the area is cleaned up. Removal is not a straightforward process of just bulldozing and burning the debris (consider the amount of plastics and toxic lawn chemicals in a typical suburban home); it requires separation of materials. Cleanup is usually done by commercial companies based on bids, with significant delays due to the need of companies to obtain sufficient information to make and provide accurate estimates. The objective is to locate and analyze the debris (e.g., area A has x cubic meters of vegetation and y cubic meters of light construction and household materials).

Direct intervention, such as manipulating valves or emergency devices, to mitigate the consequences of the event. Recent examples of the need for robots to manipulate the environment include robot submersibles inserting a tube and regulators into the damaged drill pipes at the 2010 Deepwater Horizon spill in the Gulf of Mexico and the use of robots to attempt to turn valves in the 2011 Fukushima Daiichi nuclear emergency. The objective is to provide safe, reliable action at a site while situated at a distance from the site.

1.7   Summary

Disaster robotics covers the entire disaster life cycle of *prevention, prepara-tion, response,* and *recovery.* Rescue robots are unmanned systems used as *tactical, organic, unmanned systems* that enable responders and other stake-holders *to sense and act at a distance from the site of a disaster or extreme inci-dent.* Rescue robots are usually smaller than military robots and hardened for the extreme environments and must be capable of working without GPS or wireless. Most of the land and aerial vehicles used in rescue robotics can be traced back to DARPA programs, while marine vehicle development has largely stemmed from the undersea industry.

The economic justification for robots (and associated costs of training and maintenance) is based on four effects: *saving lives, improving the long-term health and recovery of survivors, mitigation of negative effects such as envi-ronmental pollution,* and *economic recovery.* Disasters and extreme incidents occur frequently, and their impact is increasing. Robots can make a dif-ference if they are deployed in a timely fashion, though currently it takes almost a week for a robot to be used.

There are multiple stakeholders who can use robots, from federal agen-cies to municipalities and NGOs. In the United States, only one search and rescue team, New Jersey Task Force 1, and one agency, the MSHA, own rescue robots. As a result, the majority of robots used at disasters have been provided by CRASAR through CRASAR's Roboticists Without Borders pro-gram. Companies and universities who offer or bring robots risk being per-ceived as profiteers, too high risk, or too experimental if they do not have prior ties with the incident command structure.

Rescue robots are generally thought of for the immediate lifesaving response for meteorological, geological, man-made, and mining or mineral disasters. But these robots also can be used for the recovery operations and for prevention and preparation. They can perform immediately and directly engage missions such as search, reconnaissance, mapping, structural inspec-tion, in situ medical assessment and intervention, and direct intervention for mitigation. Robots can also carry out important indirect tasks such as acting as a mobile beacon or repeater, removing rubble, providing logistics support, serving as a surrogate for a team member, and adaptively shoring a collapse to make it safer for responders. Recovery operations include all of the above tasks, but robots can also assist with victim recovery and estima-tion of debris volume and types.

Often, researchers and investors focus on or are only aware of immediate lifesaving activities in a disaster, neglecting the lower-profile but high-value missions for both rescue and recovery. There is also a tendency to think in terms of replacing a person or a canine, rather than extending and supplementing those capabilities. While incidents like hazardous materials responses could benefit from robots that could do what a human would do if a person were able to survive long enough, the majority of missions will use robots for novel tasks. These formative uses of robots pose design risks, as the developer has to project how the robot would fit into larger environmental and organizational constraints.

The remainder of the book is organized as follows. The next chapter summarizes the 34 deployments of rescue robots that have been identified as of April 2013 and analyzes their performance and the lessons learned. The following three chapters discuss each of three modalities (ground, air, marine) in more detail, including typical platforms and gaps in technology. Together, chapters 1–5 provide a broad overview of the state of the practice in rescue robots. The final chapter, chapter 6, offers recommendations for working in the field with stakeholders and provides more detail about the emergency response enterprise.

2   Known Deployments and Performance

This chapter describes and analyzes each of the 34 disasters and extreme incidents where robots have been deployed. Whereas chapter 1 established the types of robots and their potential uses, this chapter considers where robots have actually been used, for what purposes, and to what degree of success. The diversity of deployments illustrates that **no one size of type of robot fits all missions.** The data in this chapter should assure agencies and emergency managers that **although robots are still far from perfect, they are useful.** The failures and gaps described should not be used as reasons to reject use of a robot but rather as decision aids in selecting a currently available robot and for proactively preparing a field team for what to expect. The chapter provides researchers, technologists, and students with a set of cross-cutting development challenges and open research questions for all of disaster robotics, noting that chapters 3, 4, and 5 will go into more depth on the challenges and questions for each robot modality (ground, aerial, and marine).

Surprisingly few deployments have been reported in the scientific or professional literature, and even fewer have been analyzed in any depth. In many cases, there is insufficient data for a detailed discussion; note that chapter 6 will go over data collection in the field, both what data are valuable and how to gather it in the hopes that future deployments will be more informative. Even if data are collected, many reports lack a unifying framework or conceptual model for analysis. Disaster robotics, in particular, and field robotics, in general, are new areas of discovery. Their newness means there is a lag in understanding how best to capture performance and even the dimensions that make up performance. Performance goes beyond simple binary declarations of mission success: it requires knowing what worked and why. This chapter expands the failure taxonomy developed by Carlson and Murphy (2005) and uses it to discuss robot performance in terms of external operational environment, malfunctions in one of the five physical subsystems of a robot (mobility, communications, sensing, control, and power), and human error.

This chapter addresses four basic questions:

• **Where have disaster robots been used (and for what)?** Disaster robots have gained media attention for their use in high-profile disasters such as the Fukushima Daiichi nuclear emergency in Japan (though ironically, Fukushima Daiichi was initially a symbol of the lack of rescue robot capability) and the World Trade Center in New York—but have they been used elsewhere? And if so, for what? Has a robot ever found a survivor?

• **How well do disaster robots work?** Given that much of the press on robots at Fukushima Daiichi highlighted robot failures (e.g., a robot not able to make it up a flight of stairs or a UAV performing an emergency landing), what is the actual performance of a rescue robot? Are they reliable enough to be used? Where are the gaps and problems? But those questions pose another question that has to be answered first: How can disaster robot performance be formally analyzed? What are the frameworks that have been used? Is there a comprehensive framework?

• **What are the general trends about disaster robotics?** Is one modality more popular than the others? Are robots used for one type of disaster more than others?

• **What are the surprises, barriers, and research gaps?** Every disaster is different and could be considered surprising, but what are the surprises in terms of mismatches between commonly held perceptions of rescue robotics and reality? What are the barriers to immediately deploying existing robots? What are the pervasive research gaps that will require sustained, long term investigations?

The list of deployments has been extracted from the scientific and response literature, news reports, and word of mouth. As a result, it is likely that some unpublicized deployments have been missed. Readers who are aware of other deployments not reported in this chapter or additional details about known deployments are asked to share their information with the author.

2.1   Where Have Disaster Robots Been Used (and for What)?

Table 2.1 shows the known history of robots used for disaster response and recovery; this table will be expanded in later chapters. Robots have been used at 29 incidents and additionally have been deployed to sites but not used five times (indicated by the shaded rows in table 2.1). In the majority of events, only one modality of robot has been used; for example, mine disasters have used only ground robots. The presence of multiple modalities does not imply coordination; for example, at the Haiti earthquake, UAVs and UMVs were used by different groups for different missions. The number(s) for each entry in the table represents the reported or inferred number of models of robot of that modality used for the event. For example, at the 2001 World Trade Center collapse, 16 different models of UGVs were on site, but only four models were deployed on the rubble pile, and two others were demonstrated in collaterally damaged buildings; thus the entry is "4." All 29 events are described later in this section.

The common question "How many lives have rescue robots saved?" implies that the only mission for a rescue robot is lifesaving; as will be seen later, this is only one of many missions of value. To date, rescue robots have not been involved in saving a specific life but have been generally successful in accomplishing search, reconnaissance, and mitigation missions. In addition, they are credited with helping to prevent additional flooding and associated loss of life and property during the 2011 Thailand monsoon season. Part of the reason why robots have not been associated with a "live save" is that they have rarely been tasked in situations with survivors or soon enough to make a difference. Robots have usually been in conditions where there was a low probability of anyone remaining alive (e.g., World Trade Center, Berkman Plaza II, missing balloonists, Pike River Mine, etc.) or the robot was deployed days or weeks after the incident (e.g., Crandall Canyon Mine), further reducing the likelihood of finding a survivor. In general, when rescue robots have been used during the rescue phase, they have been very helpful in rapidly ruling out areas to search, and responder feedback has been positive.

In all but one case, the robots had a mission or were working with or through an agency or official. As will be discussed in chapter 6, self-deployment is not only discouraged but also illegal; scientists have been arrested for using technologies at a site without permission (Ingrassia, 2007). In the cases marked with an asterisk in table 2.1, the robots were essentially being used for postdisaster experimentation and demonstrations of feasibility to the agencies rather than to provide mission-critical data. At several events, robots were initially used for the lifesaving search and rescue operations then used to assist with recovery and mitigation operations. Table 2.1 should not be interpreted as saying that there were 29 different types of UGVs deployed to disasters. In general, only a few robot models have been used; for example, the same robot, the MSHA V2 intrinsically safe (mine-permissible) version of the Remotec ANDROS Wolverine, has been used in most of the mine incidents in the United States.

Table 2.1
Robot deployments from 2001 to 2013 by modality and number of robots

Year	Event	Ground	Aerial	Marine	Year	Event	Ground	Aerial	Marine
2001	World Trade Center (USA)	4			2008	Hurricane Ike (USA)			1
2001	Jim Walter No. 5 Mine (USA)	1			2009	Historical Archive of Cologne building collapse (Germany)	2		
2002	Barrick Gold Dee Mine (USA)	1			2009	L'Aquila earthquake (Italy)*		1	
2004	Brown's Fork Mine (USA)	1			2010	Haiti earthquake (Haiti)		1	1
2004	Niigata Chuetsu earthquake (Japan)*	1			2010	Wangjialing Coal Mine (China)	1		
2004	Hurricane Charley (USA)	1			2010	Upper Big Branch Mine (USA)	1		
2004	Excel No. 3 Mine (USA)	1			2010	Deepwater Horizon (USA)			16
2005	DR No. 1 Mine (USA)	1			2010	Prospect Towers (USA)	2		
2005	McClane Canyon Mine (USA)	1			2010	Missing balloonists (Italy)			1
2005	La Conchita mudslides (USA)	1			2010	Pike River Mine (New Zealand)	2		
2005	Hurricane Katrina (USA)	1		3	2011	Christchurch earthquake (New Zealand)	1	1	
2005	Hurricane Wilma (USA)*		1	1	2011	Tohoku earthquake (Japan)	3	1	
2006	Sago Mine (USA)	1			2011	Tohoku tsunami (Japan)			9
2007	Midas Gold Mine (USA)	2			2011	Fukushima Daiichi nuclear emergency (Japan)	7	2	
2007	Crandall Canyon Mine (USA)	1			2011	Naval base explosion (Cyprus)		2	
2007	I-35 Minnesota bridge collapse (USA)			2	2011	Great Thailand Flood (Thailand)		2	
2007	Berkman Plaza II collapse (USA)	2	1		2012	Finale Emilia earthquake (Italy)	2	2	

Note: The number(s) for each entry in the table represents the reported or inferred number of models of robot of that modality used for the event. Shading means robots were present but not used.

*A deployment for purposes of demonstration/experimentation occurred after the disaster but with the permission or presence of an agency.

Most of the robot deployments will be discussed in more detail as case studies throughout the chapters that follow, but a brief overview is provided here to establish a familiarity with the incidents and deployments. All robots were teleoperated.

2.1.1   World Trade Center, New York, New York (2001)

The Center for Robot-Assisted Search and Rescue (CRASAR) deployed ground robots that had been associated with the Defense Advanced Research Projects Agency (DARPA) Tactical Mobile Robots program. Figure 2.1 shows eight of the 17 tracked and wheeled robots assembled by CRASAR from DARPA, Foster-Miller, iRobot, the University of South Florida, and the U.S. Navy Space and Naval Warfare Command San Diego. Only the three smallest and lightest models were used on the rubble pile from September 11 to 21 during the rescue phase: the Inuktun Micro-Tracks, the Inuktun Micro-VGTV, and the Foster-Miller Solem (Murphy, 2004b). The robots were used in the rubble of Towers 1 and 2 and Building 4 in an attempt to penetrate the rubble and find "shortcuts" to the basement or stairwells where firemen might have survived the collapse. The only wireless robot, the Solem, lost signal while in the rubble and was not recovered. The robots were considered successful in that they were able to penetrate far beyond what search cameras could do, and they operated in extreme heat. The robots found approximately 10 sets of remains.

When operations entered the recovery phase, the U.S. Army Tank Automotive and Armaments Command–Army Research and Development Command–Explosive Ordnance Disposal Division (USA TACOM-ARDEC-EOD) teamed up with CRASAR, and four models of robots were used from September 23 to October 2: the Inuktun Micro-Tracks, the Inuktun Micro-VGTV, the Foster-Miller Solem, and the slightly larger Foster-Miller Talon (Platt, 2002). City engineers used the Foster-Miller Solem and Talon models to travel down newly uncovered stairwells to help with inspection of the basement slurry wall. It should be noted that Platt (2002) contains numerous significant errors in describing the rescue phase (understandable as that author was only present for the recovery phase); therefore, Casper and Murphy (2003), Murphy (2003), and Murphy (2004b) should be used as sources for information about the rescue phase.

2.1.2   Barrick Gold Dee Mine, Elko, Nevada (2002)

The U.S. Mine Safety and Health Administration (MSHA) deployed a mine-permissible Remotec ANDROS Wolverine (figure 2.2) in November 2002 for a mine recovery (reopening) operation at the Storm Decline at Barrick Gold

Figure 2.1
Some of the robots deployed to the World Trade Center; only the three smallest models were used during the rescue phase.

Dee Mine in Elko, Nevada (Murphy and Shoureshi, 2008). That October, two mine rescue team members had been killed during a training exercise on mine recovery operations. The robot, named V2, was deployed from the surface down a 16-degree slope. It was able to navigate and to take continuous gas samples.

2.1.3   Browns Fork Mine, Hazard, Kentucky (2004)
The MSHA deployed V2, a mine-permissible Remotec ANDROS Wolverine, in August 2004 at the Browns Fork Mine for a mine recovery task (Murphy and Shoureshi, 2008). The robot was unsuccessful at reaching and dislodging a continuous miner (a piece of mining equipment). V2 was too tall to go further than an initial 15–20 feet into the affected area under the highwall, the working face of the coal mine. The robot was able to move some debris with its effector as a demonstration.

2.1.4   Niigata Chuetsu Earthquake, Japan (2004)
The International Rescue System Institute (IRS) inserted the Souryu serpentine robot into a house in Nagaoka City that was damaged during the

Figure 2.2
The Mine Safety and Health Administration's mine-permissible Remotec ANDROS
Wolverine, V2. (Photograph courtesy of MSHA.)

Niigata Chuetsu earthquake (Arai et al., 2008). Souryu, shown in figure 2.3,
was codeveloped with Prof. Shigeo Hirose at the Tokyo Institute of Technology and was the *first snake robot used at a disaster site*. It showed the prom-
ise of biomimetic alternatives to tracked platforms. The deployment in the
debris concentrated on demonstrating mobility to fire rescue agencies.

2.1.5   Excel No. 3 Mine, Pikesville, Kentucky (2004)
The MSHA deployed V2, a mine-permissible Remotec ANDROS Wolverine,
in December 2004 at the Alliance Resources' Partners Excel No. 3 Mine for
mine recovery after a mine fire (Murphy and Shoureshi, 2008). Because of
the 15-degree slope of the coal mine, which was layered in ice from the
use of liquid nitrogen to suppress the mine fire, the robot required a safety
rope and winch for entry and return. The robot was able to penetrate 750
feet into the mine and successfully completed the objective of providing an
assessment of the situation.

2.1.6   DR No. 1 Mine, McClure, Virginia (2005)
The MSHA deployed the V2, a mine-permissible Remotec ANDROS Wolverine, in January 2005 to the DR No. 1 Mine for mine recovery (reopening)
(Murphy and Shoureshi, 2008). The coal mine had been sealed after an
explosion that killed seven miners in 1983. The robot was able to penetrate a maximum distance of 700–800 feet into the mine with a slope of
18 degrees, and the robot arm was used move and realign ceiling supports

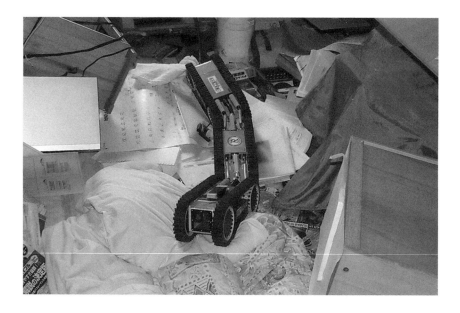

Figure 2.3
IRS Souryu serpentine robot being used at Nagaoka City after the Niigata Chuetsu
earthquake. (Photograph courtesy of IRS.)

in order to progress into the mine. The robot executed two runs: during run
1, a fiber-optics failure occurred, while run 2 was successful.

2.1.7   McClane Canyon Mine, Grand Junction, Colorado (2005)
The MSHA deployed the V2, a mine-permissible Remotec ANDROS Wolverine, in November 2005 to the McClane Canyon Mine for mine recovery
(Murphy and Shoureshi, 2008). Trials were conducted to establish manipulation capabilities at the coal mine. In this case, the robot was tasked to
close five doors and pull out timbers holding up a mine fan. The robot was
generally unsuccessful, and eventually the fiber-optic control cable was cut.

2.1.8   La Conchita Mudslides, La Conchita, California (2005)
CRASAR deployed two Inuktun VGTV Xtreme robots, a waterproof upgrade
of the tracked Inuktun Micro-VGTV used extensively at the World Trade
Center, to assist with the search for missing residents (Murphy and Stover,
2006b; Murphy and Stover, 2008). Both robots detracked within minutes
of deployment and were unsuccessful in investigating collaterally damaged houses.

2.1.9 Hurricane Katrina, Mississippi Coast (2005)

The Florida State Emergency Response Team deployed two UAVs and Florida Task Urban Search and Rescue Task Force 3 deployed one UGV in Mississippi under a mutual emergency aid compact to assist with rescue operations after Hurricane Katrina, and then CRASAR returned to document the disaster. This was the *first deployment of small UAVs for a response* and their first use at Hurricane Katrina, though other UAVs were used later in the response as part of military aid. CRASAR, a member of SERT, deployed two types of small unmanned aerial vehicles to Hurricane Katrina in September 2005 to assist with the rescue phase (Murphy, 2006; Murphy et al., 2006c), and the deployment was the first use of UAVs for a disaster response. A small, fixed-wing UAV, not named at the request of the contributing agency but later identified as an AeroVironment Raven, and an iSENSYS T-Rex variant miniature helicopter (figure 2.4) flew two successful missions to determine whether people were stranded in the area around Pearlington, Mississippi, and if the cresting Pearl River was posing immediate threats. Later, CRASAR returned with an iSENSYS IP-3 miniature helicopter specifically designed for structural inspection; the 32 flights successfully examined structural damage at seven multistory buildings (Pratt et al., 2006; Pratt et al., 2009). Florida Urban Search and Rescue Task Force 3 deployed an Inuktun VGTV Xtreme to search two apartment buildings in Biloxi, Mississippi, where it was unsafe for response (Micire, 2008).

2.1.10 Hurricane Wilma, Ft. Myers, Florida (2005)

CRASAR deployed an AEOS marine USV (figure 2.5) and an iSENSYS T-Rex variant miniature helicopter in October 2005 to assist with the recovery phase of Hurricane Wilma (Murphy et al., 2008a). The USV used an acoustic

Figure 2.4
(Left) Fixed-wing UAV and (right) T-Rex VTOL UAV used at Hurricane Katrina.

camera to inspect seawalls, docks, and harbor channels at Ft. Myers as well as the damaged fishing pier at the base of the SR951 bridge. The missions were successful, finding evidence of scour in the subsurface dock footings and debris in some channels. CRASAR also experimented with cooperative UAV–USV operations (Murphy et al., 2006d).

2.1.11   Sago Mine, Sago, West Virginia (2006)

The MSHA deployed the V2, a mine-permissible Remotec ANDROS Wolverine, in January 2006 to the Sago Mine (Murphy and Shoureshi, 2008; Murphy et al., 2009a). The objective was to assist with the rescue efforts for 13 missing miners. After moving approximately 600 feet ahead of rescuers, the teleoperated robot ran off the rail and over a 30-inch drop to the floor of the coal mine, damaging the robot beyond use.

2.1.12   Midas Gold Mine, Midas, Nevada (2007)

The Naval Air Station Fallon bomb squad team deployed an Allen-Vanguard in June 2007 followed by CRASAR deploying an updated Inuktun VGTV Xtreme (an improvement to the Xtreme used at the La Conchita mudslides)

Figure 2.5
AEOS-1 USV inspecting a bridge and piers after Hurricane Wilma.

to the Midas Gold Mine collapse to search for a missing miner, presumed dead at the time (Murphy and Shoureshi, 2008; Murphy et al., 2009a). The Allen-Vanguard was successful in getting pictures of the loader the miner had been operating when the floor collapsed, but it was too heavy to be lowered more than 70 feet into the vertical void and damaged ledges where a body might be trapped. The Xtreme was able to penetrate 120 feet into the void but was only allowed to search the main cavity while suspended 30 feet from the floor because of other search operations. The robot targeted the area where the body was recovered but did not have sufficient lighting to capture images of the body.

2.1.13   I-35 Bridge Collapse, Minneapolis, Minnesota (2007)
The FBI Evidence Response Team deployed two unnamed ROVs in August 2007 to the collapse of the I-35 bridge in Minneapolis, Minnesota (Goldwert, 2007). The larger of the two robots was reportedly too large to maneuver in the submerged wreckage, while the smaller, a Video Ray based on photographs (FBI, 2007), was apparently helpful in documenting the site and locating victims trapped under the wreckage.

2.1.14   Crandall Canyon Mine, Huntington, Utah (2007)
The MSHA deployed a custom Inuktun Mine Cavern Crawler robot in August 2007 to the Crandall Canyon Mine to search for six missing miners (Murphy and Shoureshi, 2008; Murphy et al., 2009a). The robot had to travel through an irregular borehole with a nominal 8⅞-inch diameter (figure 2.6). Four attempts were made to enter the mine through two different boreholes, but only one was successful. In the fourth run, the robot was able to travel more than 1,400 feet through the borehole then search about 7 feet on the mine floor, which was largely impassable due to the debris and drilling tailings. The robot provided no sign of the miners.

2.1.15   Berkman Plaza II Collapse, Jacksonville, Florida (2007)
CRASAR deployed an Inuktun VGTV in December 2007 to the collapse of the Berkman Plaza II parking garage to assist in the search for a missing worker, and then with the IRS deployed a Active Scope Camera ground robot (figure 2.7), an Inuktun VGTV tracked robot, and an iSENSYS IP-3 miniature helicopter to assist with the recovery (structural forensic inspection) (Pratt et al., 2008). The Inuktun VGTV was successfully able to search one void behind a "widow maker" dangerous overhang where there was a high probability of the missing worker and see that he was not there. The caterpillar-like Active Scope Camera was successful in helping document

Figure 2.6
Inuktun Mine Cavern Crawler being inserted into borehole at the Crandall Canyon Mine response.

the internal state of the pancake portion of the collapse, operating in spaces of less than 2 inches (Tadokoro et al., 2009). The Inuktun VGTV was successful in providing conclusive documentation of structural defects in the standing portion of the parking garage. The IP-3 provided views that manned helicopters had been unable to obtain due to kicking up of dust and inability to fly close to structures (Pratt et al., 2008). One emergency landing of the IP-3 occurred due to tangling of the tether.

2.1.16   Hurricane Ike, Rollover Pass, Texas (2008)
CRASAR deployed an AEOS USV, a Video Ray ROV, and a YSI Oceanmapper AUV in December 2008 to the Rollover Pass Bridge to assist with the recovery (bridge inspection) after Hurricane Ike (Murphy et al., 2009c; Murphy et al., 2011b). The USV used an acoustic camera successfully to examine the bridge pilings for erosion indicating a possible collapse and successfully mapped the upstream area, showing that there was a low likelihood of debris being swept against the bridge. The ROV was unable to provide useful data because of the turbidity of the water, and its tether became tangled

Figure 2.7
IRS Active Scope Camera caterpillar-like robot at the Berkman Plaza II collapse.

in the debris. While on the surface, the AUV was unable to navigate close to the bridge because of the loss of GPS signal due to "urban shadows."

2.1.17   L'Aquila Earthquake, L'Aquila, Italy (2009)
Members of the Cognitive Cooperative Robotics Laboratory at the Sapienza University of Rome flew their custom Ascending Technologies quadrotor UAV in the aftermath of the L'Aquila earthquake for the L'Aquila Fire Department (figure 2.8). The UAV was extremely small and is being evolved to fly into the interiors of buildings. The deployment in the debris concentrated on demonstrating mobility to fire rescue agencies.

2.1.18   Haiti Earthquake (2010)
Two different robots were used for separate missions in January 2010. The U.S. Navy and U.S. Army joint service Mobile Diving and Salvage Unit 2 (MDSU 2) deployed SeaBotix LBV ROVs to identify debris in shipping lanes for divers to clear safely, thereby permitting offloading of humanitarian aid (Lussier, 2010). The ROVs were equipped with a Gemini acoustic camera and a video camera. Evergreen Unmanned Systems deployed an Elbit

Skylark to survey the state of orphanages near Leogane. The Skylark has autonomous waypoint navigation capability, but it was not used.

2.1.19 Deepwater Horizon Oil Spill, Gulf of Mexico (2010)

British Petroleum (BP) deployed between 14 and 16 ROVs to assist with inspection of an underwater oil spill and to insert mitigation devices during April 2010. The Schilling Ultra Heavy Duty, Oceaneering Maxximum and Millenium ROVs, which are commonly used on drilling rigs, were directly mentioned, though there may have been other models (Newman, 2010). Activities included initial manipulation of the blowout preventer to attempt to activate it, cutting of the riser pipe, insertion of a collection tube into the riser, and installation of a lower marine riser package (LMRP) cap to better collect oil while "kill" operations continued. There were at least four notable events with the robots. An ROV collided with pipework dislodging a tube inserted into the leaking blowout preventer (Chazan, 2010). An ROV apparently collided with the LMRP cap, causing the LMRP cap to be removed, brought to the surface, repaired, and then replaced (Newman, 2010). Finally, an ROV may have collided with another ROV (Bob99Z, 2010).

Figure 2.8
Sapienza University of Rome custom quadrotor UAV at L'Aquila earthquake.

Figure 2.9
Two members of New Jersey Task Force 1 using an Inuktun VGTV Xtreme (not visible) at the Prospect Towers parking garage collapse. (Photograph courtesy of New Jersey Task Force 1.)

2.1.20   Prospect Towers Collapse, Hackensack, New Jersey (2010)
The Newark Fire Department's Urban Area Security Initiative (UASI) Strike Team deployed an Inuktun VersaTrax 100 and New Jersey Task Force 1 deployed an Inuktun VGTV Xtreme to explore the collapse of the parking garage of the Prospect Towers condominium complex in July 2010 (figure 2.9). The shoebox-sized robots were used to see vehicle information, license plates, and vehicle makes and models in areas not safe for responders. The robot in figure 2.9 is being used to examine a car that was the same model as that of a family unaccounted for and might be trapped. By being able to read the license plate, the responders could tell this wasn't their car.

2.1.21   Missing Balloonists, Adriatic Sea (2010)
The Italian Coast Guard deployed an unknown UMV (possibly a SeaBotix LBV) in October 2010 to search for bodies and debris near Vieste, Italy, from the crash of the Abruzzo-Davis entry in the Gordon Bennett Gas Balloon Race. The UMV worked in areas identified from an aircraft reconnoiter and operated at depths of 200 meters but did not find the victims (Rizzo, 2010).

2.1.22   Pike River Mine, Greymouth, New Zealand (2010)

The Tasman Police District deployed two unknown New Zealand Defense Force bomb squad robots in November 2010 and had a Western Australia Water Company pipeline inspection robot on site and the MSHA's V2 mine-permissible robot en route to understand the state of the coal mine and to find paths to the victims. The mine was unsafe for human entry. The first robot failed after reaching 550 meters into the mine because of falling water (TVNZ, 2010a; TVNZ, 2010b) but was successfully restarted and moved out of the way of the second robot before running out of battery power. The first and second robots were presumed destroyed in a second explosion that ended any expectations of survivors (TVNZ, 2010c).

2.1.23   Christchurch Earthquake, New Zealand (2011)

On February 22, 2011, the Christchurch earthquake struck New Zealand. One of the damaged buildings, a Roman Catholic cathedral, was inspected by structural engineers from Opus International Consultants (Lester, Brown, and Ingham, 2012). A consumer Parrot AR drone was initially used to fly in a window. An iRobot Packbot from the New Zealand Army was then sent in on June 13, 2011, to conduct further inspections, with an Army operator working side by side with the structural engineers. The data from the UGV gave the first clear view of the interior since the quake and showed extensive damage, allowing appropriate shoring and remediation.

2.1.24   Tohoku Earthquake, Japan (2011)

Both ground and marine vehicles were used in the response to the Tohoku earthquake and tsunami, which occurred in Japan on March 11, 2011. The earthquake damage to multistory buildings was slight, but in two instances the IRS fielded ground robots. IRS inspected a partially collapsed building at the Hachinohe Institute of Technology with KOHGA3, but in general multistory commercial buildings survived the earthquake without a full collapse or loss of life. Two IRS research ground robots, Kenaf and Quince, were used with a Pelican UAV from the University of Pennsylvania to test multirobot collaborative mapping in a damaged building at Tohoku University (Michael et al., 2012).

2.1.25   Tohoku Tsunami, Japan (2011)

Marine vehicles were more extensively used in the response to and recovery from the Tohoku earthquake and tsunami, which occurred in Japan on March 11, 2011. CRASAR deployed four ROVs and a five-person team in April 2011 to support recovery operations with the IRS at the request of the Minamisanriku and Rikuzentakata townships. (The Port of Hachinohe had requested ROV assets on March 13, but travel restrictions due to the Fukushima Daiichi incident led to travel delays.) A SeaBotix SARbot with computer-enhanced video and a Trident Gemini acoustic camera were used to inspect critical infrastructure and to recover bodies, while the Seamor ROV with a DIDSON acoustic camera and a miniature Access AC-ROV were used less so. The SeaBotix LBV served as a backup to the SARbot and was not needed. The IRS–CRASAR team was credited with reopening the Port of Minamisanriku (NewsCore, 2011). Prof. Tamaki Ura from Tokyo University fielded an unnamed ROV, which recovered two dead bodies from an undisclosed location (Ferreira, 2011). Prof. Shigeo Hirose at Tokyo Institute of Technology also created an experimental ROV and deployed with the Japanese Ground Self-Defense Forces.

CRASAR returned in October to continue joint operations with IRS using a YSI Echomapper and SeaBotix SARbot, though the YSI Echomapper malfunctioned and could not be used. IDEA Consultants (Japan) joined the operations with their Mitsui ROV. Other private industries also began deploying ROVs. Shibuya Diving Industry bought and used a Video Ray Pro 4, though the extent of its use is not known (Shibuya, 2012).

2.1.26   Fukushima Daiichi Nuclear Plant, Fukushima, Japan (2011)

The exact count of robots used for the emergency at the Fukushima Daiichi nuclear power plant is unclear at this time in part because of lack of information from the owner, TEPCO. It appears that at least two UAVs and seven types of UGVs were used through July 2011. A Honeywell T-Hawk brought in by Westinghouse, a TEPCO subcontractor, appears to be the first robot officially used at Fukushima Daiichi with flights starting on April 10, 2011, nearly a month after the March 12, 2011, series of failures that led to the meltdown. However, apparently independently of TEPCO or any airspace permissions, the Air Photo Service Company of Japan flew a custom fixed-wing UAV and took pictures of the facility on March 20, 2011. Shortly thereafter, two QinetiQ Talons were used in conjunction with Bobcat construction loaders that had been converted for teleoperation to clear debris and radioactive waste, allowing access to the site (Kawatsuma, Fukushima, and Okada, 2012). A BROKK90 UGV, two BROKK330D UGVs, and one BROKK800D UGV were used to clear radioactive debris in the reactor buildings (Kawatsuma, Fukushima, and Okada, 2012). iRobot Packbots were the first robots in the reactor buildings and were used to read gauges and conduct general reconnaissance. At the end of April, a box truck driven by two operators in protective gear deployed a gamma-radiation camera from the truck bed and a QinetiQ Talon precisely to identify hot spots within the site (Ohno et al., 2011). In June, the IRS Quince robot was deployed to survey the reactor building interiors (Nagatani et al., 2011), as the iRobot Packbot had difficulty ascending three flights of muddy stairs. More teleoperated construction equipment such as backhoes was brought in (Boyd, 2011). In July, an improvised robot vacuum cleaner using an iRobot Warrior base was used to clean up radioactive concrete dust, and a JAEA-3 custom robot was used for reconnaissance (Kawatsuma, Fukushima, and Okada, 2012). The JAEA-3 was based on the RESQ-A robot developed by the Japan Atomic Energy Agency after a 1999 nuclear incident.

2.1.27   Evangelos Florakis Naval Base Explosion, Cyprus (2011)
On July 11, 2011, an explosion occurred at the Evangelos Florakis naval base in Cyprus killing 13 people and damaging the adjacent Vasilikos Power Plant, which is responsible for a significant portion of the island's electricity. A team from the German Aerospace Center was dispatched through the European Civil Protection Mechanism (Angermann, Frassl, and Lichtenstern, 2012). On July 26, 2011, the team used an AscTec Falcon and an AscTec Hummingbird to inspect the damage and create a three-dimensional image of the power plant.

2.1.28   Great Thailand Flood (2011)
The Siam UAV Industries company worked with authorities for 45 days during the Great Thailand Flood (March–April 2011) to prevent damage to Bangkok (Srivaree-Ratana, 2012). At least two UAVs, presumed to be fixed-wing, assisted in collecting data for flood simulations.

2.1.29   Finale Emilia Earthquake, Italy (2012)
On May 29, 2012, a series of earthquakes struck northern Italy destroying historic churches at Mirandola, Italy. As part of the recovery effort, a team of university researchers from the European Union–funded Natural Human–Robot Cooperation in Dynamic Environments Project (NIFTi) fielded two UAVs and two UGVs under the direction of the Italian National Fire Corps. The robots were used from July 24 to July 29 to inspect the exteriors and interiors of two churches that had not been entered because of safety reasons. The robots were successful and provided engineers and cultural historians with information that could not be obtained otherwise.

2.1.30   Disasters Where Robots Were Deployed but Not Used

It is illuminating to consider the five cases where robots were deployed but not used. There was one such case in 2001, one in 2004, one in 2009, and two in 2010: Jim Walter No. 5 Mine explosion (USA), Hurricane Charley (USA), Historical Archive of Cologne building collapse (Germany), Wangjialing Coal Mine (China), and Upper Big Branch Mine (USA), respectively. In the Wangjialing Coal Mine and Upper Big Branch Mine incidents, robots were present but not used because of presence of water or debris. This illustrates how the extreme environments for both the robots and the operators distinguish this class of applications from routine inspection or even improvised explosive device (IED)-related tasks.

On September 24, 2001, explosions at the Jim Walter No. 5 Mine killed 13 miners. The MSHA deployed a Remotec ANDROS Wolverine (Murphy and Shoureshi, 2008). However, the robot was not inserted, as it was not mine permissible and could ignite the methane in the coal mine. An attempt to make the robot mine permissible by flooding the platform's compartments with inert gases was deemed unacceptable by safety officials, supporting the observation that rescue robots are not constructed "on the fly."

On August 13, 2004, Hurricane Charley intensified to a category 4 storm and made landfall near Punta Gorda, Florida, becoming the seventh most costly catastrophe and the fifth most expensive hurricane in U.S. history according to the National Hurricane Center. CRASAR deployed as part of Florida Task Force 3, the first team to enter Punta Gorda, which occurred around midnight (Murphy, Stover, and Choset, 2005). The team brought an Inuktun VGTV and a set of sensors, including a new laser illumination system. As will be discussed in chapter 3, UGVs are not suited for rapid inspection of residential housing, and the robot was not used.

On March 3, 2009, the Cologne, Germany, Historical Archive building and several others collapsed due to excavation of a new subway route. CRASAR deployed the Inuktun VGTV Xtreme with the IRS's Active Scope Camera robot. The original intent was to search the Historical Archive building, which had suitable voids for the two robots. However, once CRASAR and IRS were on site, the mission shifted to search dense rubble from a brick residential structure instead. The density of the rubble eliminated the VGTV Xtreme from consideration, while the unsafe conditions for the Active Scope Camera operator eliminated use of that robot (Linder et al., 2010).

On March 28, 2010, the Wangjialing Coal Mine in China flooded, trapping 115 miners. An unknown UGV (probably similar to a Remotec

ANDROS class of robots) of unknown provenance was reported at the site on April 1, 2010, for search (Low, 2010). However, it was not used, probably either because of the presence of water (assuming the robot was not water-proof) or because it was too big for the void space. The miners were eventually found by the drilling of boreholes, though 38 were reported dead.

On April 5, 2010, an explosion at the Upper Big Branch Mine killed 25 men and left four unaccounted for. The MSHA immediately deployed their nearby V2 robot, a mine-permissible (explosion-proof) variant of the Remotec ANDROS Wolverine. The intent was to enter via the surface entry, where the robot could search more quickly for the missing miners than could human responders encumbered by safety gear. However, the presence of significant debris and restricted passages around mining equipment meant the robot was unlikely to succeed, and it was not used (Statement Under Oath of Virgil Brown, 2011).

2.1.31 Disasters Where Robots Were Requested but Not Available

In addition to the deployments shown in table 2.1, there are two cases where robots were requested but a suitable robot was not available: the 2010 San Jose Copper–Gold Mine in Chile and the 2011 San Juan De Sabinas Coal Mine explosion in Mexico. These illustrate the extreme environmental conditions for robots.

In the August 5, 2010, San Jose Copper–Gold Mine incident, robots were requested but none would fit the size restrictions. On August 10, 2010, Parsons Brinkerhoff asked CRASAR for robots capable of assisting in the search for 33 missing miners. The San Jose mine operators wanted to deploy a robot down an irregular borehole and have the robot travel on the floor of the mine to go beyond what a fixed borehole camera could see. This was conceptually similar to the robot deployment at the 2007 Crandall Canyon Mine collapse where a team from PipeEye International, Inuktun, and CRASAR and led by MSHA deployed a custom Inuktun Mine Cavern Crawler with a range of 1,700 meters in 22-centimeter-diameter boreholes approximately 425 meters deep. Unfortunately, the San Jose boreholes were 8 centimeters in diameter and more than 700 meters deep, far too small for the Mine Cavern Crawler and too small or too long for existing or rapidly customizable robot technology. CMU and the IRS in Japan were later contacted, and they forwarded the requests back to CRASAR.

On May 3, 2011, an explosion rocked the San Juan De Sabinas Coal Mine killing 14 miners. The MSHA offered the use of the V2, the only mine-permissible robot in the world, but the mine was not a good fit for the Wolverine-variant robot, as the miners were in a small area blocked by rubble.

2.2   How Well Do Disaster Robots Work?

Asking how well do robots actually work cuts to the heart of technology and deployment issues. Engineers, computer scientists, and other technologists want to know what is currently not working so that they can conduct effective research and development. Stakeholders such as agencies and emergency managers want to understand the risk in using robots and get a better sense of how effective they are. Given that our studies show **a UGV fails 10 times as frequently as the same robot in a laboratory setting** and that **UGVs have a mean time between failure (MTBF) in the field of 6–20 hours** (Carlson and Murphy, 2005), there is clearly a gap in understanding of the real world. Although chapters 3, 4, and 5 will describe the performance of each modality of robot and provide case studies, this section provides a cumulative overview independent of modality.

The performance of robots is quite good, considering the extremities of the environments and the nascent robot technology. Robots encountered significant problems that terminated their run in only about half of the incidents (13 of 24), and only a relatively small number of robots, five, have been lost. In three incidents, a breakdown of a robot interfered with the response or made things worse, but each disruption was temporary: A malfunctioning robot temporarily blocked other robots and responders from entering the Pike River Mine, and ROVs at the Deepwater Horizon incident collided with mitigation systems twice, allowing the oil leak to temporarily resume. In contrast, robots have sometimes exceeded expectations; for example, the Xtreme at the Midas Gold Mine continued to perform well despite boulders breaking off pieces of the track guides.

Another aspect of how well robots worked is whether they were able directly to meet the mission objectives or whether the robot or the operating protocol for using the robot had to be modified; that is, a workaround was required. Workaround is a term used in cognitive science to describe how people adapt to glitches, complexity, or other problems when using a technology for a desired purpose (Woods and Hollnagel, 2006). A workaround stems from a mismatch in the design of the robot with the application: It is similar to a nonterminal failure in that the robot can still be used but is different in that it reflects a design problem. An example of a workaround for a physical mismatch of the robot to the mission was the tying of a radiation sensor to the outside of a Honeywell T-Hawk UAV at the Fukushima Daiichi response in order to get an aerial radiological survey. In this case, the robot did not have the flexibility to add new payloads and transmit their data. An example of a workaround for operations is the practice of

using two UGVs to enable one UGV to navigate through extremely tortuous passages or for manipulation tasks, such as at Fukushima Daiichi where
one robot provided an external view of the second robot's manipulator arm
in order to speed up opening of doors.

This section looks at both (i) the known cases where rescue robots failed
and discusses the causes in terms of external factors such as the environment, breakage in one of the five physical subsystems of the robot, or
human error and (ii) workarounds. The biggest source of rescue robot failures is "human error": 10 of the total 20 failures. This should be no surprise
given our analysis of UGV failures that reported a robot had a physical
failure on average once every 24 hours of use, and human failures occurred
once every 17 minutes of use (Carlson, Murphy, and Nelson, 2004). It
should be noted that human error appears to be in large part related to
the fairly primitive robot designs and manufacturing limitations associated
with emerging technologies rather than to operator deficiencies. Wireless
communications failures have not been a significant problem because all
but two models of UGVs have been tethered. The relatively few losses of
the robot or problems with power indirectly suggest that responders are
conservative in how they deploy robots—they don't risk the robot.

2.2.1   Failure Taxonomy

To examine rescue robot failures, a framework or taxonomy is needed to
aid group performance. Figure 2.10 shows a generalization of the failure
taxonomy we created (Carlson and Murphy, 2005) to categorize UGV failures; this taxonomy will be used to discuss rescue robot failures regardless
of modality. A failure either terminates a run or a mission (*terminal failure*)
or reduces the capability of the robot (*nonterminal failure*). This chapter will
only discuss terminal failures as there are data available for those, but nonterminal failures are less consistently reported.

A failure of a robot stems from

• an **external factor in the environment**, such as an explosion or other
quality of the environment;
• **breakage or a malfunction in one of the five physical subsystems** of
the robot (mobility of an effector either for navigation or manipulation,
communications, sensing, control systems, or power); or
• **human error**, such as driving a robot off a path, which is often labeled as
a lack of situation awareness.

The taxonomy does not capture the cause of the failure. There are many
potential causes for a failure. Ascribing a cause can be difficult; for example,

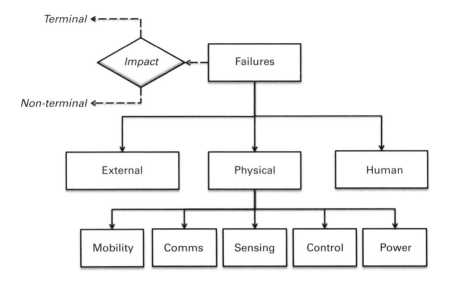

Figure 2.10
Robot failure taxonomy.

if a robot has a short circuit of its control system because it is exposed to water, is the root cause the environment, a design flaw by the human creator, or an error by the operator who drove the robot?

2.2.2 Overview of Rescue Robot Terminal Failures

Table 2.2 shows that robots had 27 terminal failures at 13 incidents, or at almost half of the total 29 incidents where robots were used. The data also suggest that some combinations of robots and incidents were tougher than others, as the 13 incidents that did have a failure had an average of 1.5 failures. In two of the 20 terminated runs, the robot failed on its return. In five cases, the robot was lost and not recovered. In two incidents, the cause of the failure is unknown.

Two robots seemed to be particularly unfortunate, illustrating the malapropism "snatching defeat from the jaws of victory." The Mine Cavern Crawler at the Crandall Canyon Mine collapse had to be removed to clean the sensors before it completely passed through the borehole into the mine. Once it entered the mine, it got stuck on reentry into the borehole. Finally, after operators successfully guided it back in the borehole, a cave-in trapped the robot on its ascent, and the robot was lost. The first New Zealand Defense Force robot used at the Pike River Mine explosion entered the mine, experienced a control failure due to water seeping from the mine

Table 2.2
Terminal failures by incident and type

Incident	Robot	Type UGV	UAV	UMV	Lost?	External Factor	Mobility	Comms	Sensing	Control	Power	Human
World Trade Center	Solem	X			X			Wireless signal lost				
	Talon	X										
	VGTV	X					Detracked (heat)			Sudden loss of control		Stuck
La Conchita mudslides	Xtreme	X					Detracked (mud)					
	Xtreme	X					Detracked (carpet)					
Excel No. 3 Mine	V2	X										Tether severed
DR No. 1 Mine	V2	X										Tether severed
McClane Canyon Mine	V2	X										Tether severed
Sago Mine	V2	X										Crashed
Berkman Plaza II	iSENSYS		X							Unknown		
Crandall Canyon Mine	Inuktun Cavern Crawler	X			X	Cave-in			Fouling due to mud, foam			Stuck

Table 2.2 (continued)

Incident	Robot	Type			Lost?	External Factor	Robot Physical Subsystem					
		UGV	UAV	UMV			Mobility	Comms	Sensing	Control	Power	Human
Pike River Mine	NZDF No. 1	X			X	Explosion				Short-circuit	loss of power	
	NZDF No. 2	X			X	Explosion						
Hurricane Ike	Video Ray			X								Tether tangled
Deepwater Horizon	Unidentified			X								Collision with pipe
	Unidentified			X								Collision with collection cap
	Unidentified			X								Collision with another ROV
Tohoku tsunami	SARbot			X								Tether tangled
	YSI Echomapper			X					Would not sense properly			
Fukushima Daiichi	T-Hawk		X		X		Unknown					
	iRobot Packbot	X					Could not climb stairs		Humidity fogged cameras			
	Quince	X			X	Facility drawings wrong						Tether cut

NZDF, New Zealand Defense Force.

roof, was successfully restarted a few hours later, and then ran out of power before being destroyed by the second explosion.

2.2.3   Terminations Due to the Environment

The environment was only responsible for four failures, though in three of those the robot was lost. The uncased borehole at the Crandall Canyon Mine eroded and caved in, thus trapping the Inuktun Mine Cavern Crawler UGV about 50 feet from the surface. The location of the borehole on the mountain prevented any practical means of extrication, and after several days of trying, the tether eventually snapped. The methane-rich atmosphere in the Pike River Mine led to an explosion that destroyed the two New Zealand Defense Force UGVs. At the Fukushima Daiichi nuclear emergency, the IRS Quince UGV entered the reactor building and climbed a flight of stairs only to discover that the landing was smaller than shown on a drawing, making it physically impossible for the robot to proceed; the mission was terminated.

2.2.4   Terminations Due to Subsystem Failures

Eleven of the failures with a known cause are associated with robot subsystems, though some manufacturers argue that these are due to human error. Mobility has the largest number of failures (four), followed by sensors (three), control (two), then communications and power tied with one failure each.

Three of the four *mobility* failures are due to detracking on the tank-like Inukun VGTV and Xtreme. In all cases, the robots were recovered, though in one case without the track. The Inuktun VGTV and its waterproof Xtreme sibling both use an open track system. This is an understood design trade-off between the risk of something getting between the tread and the pulleys (hence the term "open track") versus the ability to lift itself up to help climb over obstacles or see things. UGVs such as the iRobot Packbot use flippers with closed tracks to achieve a similar variable geometry. At the World Trade Center response, an Inuktun VGTV detracked and had to be pulled out and serviced. At the La Conchita mudslides, the Xtreme detracked on both of its two runs. One run was on sticky mud with vegetation under a house while the second was in a house on shag carpet. The iRobot Packbot initially used at the Fukushima Daiichi nuclear accident was unable to climb the wet, steep metal stairs.

Sensors account for three terminal failures. At the Crandall Canyon Mine, the Inuktun Mine Cavern Crawler had to be physically removed, its cameras cleaned, and then reinserted because groundwater, mud, and drilling foam coated the robot's cameras. **Sensor fouling** from humidity fogged the cameras on a Packbot at Fukushima Daiichi causing the termination of the run. The Doppler Velocity navigation sensor on the YSI Echomapper AUV at the IRS–CRASAR Tohoku tsunami response malfunctioned on the first run and could not be repaired.

Control failures account for two and *power* for one of the remaining terminal failures from a subsystem. In one case, the iSENSYS IP-3 UAV began behaving erratically, and the operator landed it. The New Zealand Defense Force robot no. 1 appears to have been a magnet for trouble. The waterproof robot was damaged by water seepage from the mine roof but was successfully restarted and moved out of the way before running out of battery power. It was presumed destroyed in an explosion.

Communication losses accounted for one terminal failure. The first rescue robot to be lost and not recovered was the first wireless robot to be used: the Solem or mini-Talon during the World Trade Center response. The robot, the only wireless UGV used during the initial response phase, was returning from searching a void when apparently the wireless communications antennas were grounded on the upper surface of the void. The safety line broke, and the robot was never recovered from the rubble. Wireless robots were not used for the remainder of the high-profile rescue phase but were used during the later recovery phase at the World Trade Center—with safety lines made from the same cables used to arrest jet landings on aircraft carriers!

2.2.5 Terminations Due to Human Error

Twelve runs were terminated (two only temporarily) because of human error manifested in four ways: communication tethers being run over and cut, crashes, collisions, and tether tangling. The designation of the source of the problem as "human error" is not entirely fair, as in several cases it appears that outdated interfaces and protocols were being used or available intelligent assistance was not included. In these cases, the human error might be the designer's, not the operator's, responsibility. Human error will be discussed further in chapter 5.

Tethers were severed at three mine collapses and at the Fukushima Daiichi nuclear accident. The MSHA V2 severed the fiber-optic cable at the DR No. 1 Mine and McClane Canyon Mine responses. Despite a self-controlling spool, the fiber-optic cable has some "extra" line played out, which is susceptible to being run over and crushed or cut as the operator attempts to back up the robot in order to move through narrow spaces. In each case, the robot was eventually recovered by mine rescue teams. The operator was expected to remember the location of the tether and avoid it. The Quince robot was lost at Fukushima Daiichi when the communications tether was snapped as the robot navigated through a stairwell.

The MSHA V2 crashed and fell over during the Sago Mine collapse response, experiencing significant damage. The robot had been following a narrow-gauge railway when the operator did not see that the robot was drifting to one side. Normally, going outside a rail would have no consequence as the rails are nearly flush with the mine floor, but in this case, the rails had narrowed to become a short bridge. When the robot ran off the rail, it dropped more than 30 inches to the mine floor and rolled, damaging the robot beyond use. Fortunately, it fell out of the way of rescue teams so the failure did not block the passage the way the robot at the Pike River Mine response did. The operator was expected to navigate precisely (i.e., maintain a narrow focus) while actively surveying the area (i.e., maintain a broad focus of attention).

The Deepwater Horizon incident had two, possibly three, collisions. In two of the collisions, the mitigation efforts had to be temporarily halted. In one, the ROV collided with the pipework, dislodging the tube inserted into the leaking blowout preventer and collecting some of the leaking oil (Chazan, 2010). In the second case, an ROV apparently collided with the LMRP cap, causing the LMRP cap to be removed, brought to the surface, repaired, and then replaced (Newman, 2010). A third report suggests that an ROV may have collided with another ROV (Bob99Z, 2010), and it presumed that this at least temporarily interfered with the mission. As the ROVs were teleoperated, the operators were responsible for maintaining a sense of the surroundings of the robots.

ROVs encountered problems as well. The tether for the Video Ray ROV used for evaluating the Rollover Pass Bridge after Hurricane Ike became wrapped around a pipeline. A similar situation occurred at a run with the SeaBotix SARbot that was being used for port clearing at Minamisanriku; it got wrapped around a mooring line and was recovered The operators were unable to see the tether in the murky water and could not maintain an understanding of where the robot was relative to underwater structures. The tether tangling at Rollover Pass Bridge was exacerbated by the swift current and high turbidity of the water.

In the two cases where UGVs got stuck, both were resolved by the operator taking a long break or sleeping on it, then trying again with a fresh start. This again indicates the type of intense cognitive load that the operators work under. The Foster-Miller Talon UGV became wedged in a void as it was exiting the last run for the entire CRASAR operation at the World Trade Center and was eventually extracted later that day by a combination of pulling on the safety tether and the operator's choice of movements.

The designation of the source of the problem as "human error" is a bit unfair for at least two reasons. First, the systems are not well designed for teleoperation, in part because human–robot interaction (HRI) is a new field. HRI improvements are, naturally, lagging deployments, though some manufacturers are not proactive in adopting results. Our HRI studies (Burke and Murphy, 2004) have shown that performance is much higher with two operators working together (one to drive, one to look) than it is for the single operator used by MSHA. Indeed, taken together, many HRI studies including ours suggest that operator interfaces put too much raw data on the visual display, both distracting the operator and increasing cognitive workload needed to interpret what it means. For example, the Inuktun Xtreme has a display of the force on each track, which can be used to infer incipient detracking. This is clearly an improvement over no information; however, an expert operator at the La Conchita mudslides could not notice it and react fast enough. Thus, the detracking was tallied as a mobility failure, not a human operator failure.

The second reason that the human error categorization may be unfair is that robots are often sensory and algorithmically deprived. Until recently, most UGVs had only a forward-facing camera (and often one with insufficient visual acuity as the NIST Response Robot Exercises are showing) with no way of seeing behind. But ROVs are currently in that deprived state where the sensors all face forward. Many robots do not take advantage of miniature range sensors or algorithms such as probabilistic reasoning about occupied space to avoid collisions.

2.2.6   Nonterminal Failures and Workarounds

Analyzing nonterminal failures and workarounds is harder than for terminal failures. Published reports reflect a possibly unconscious bias to present robots in a purely favorable light, which undermines the knowledge of what should be improved. The data needed for an analysis often require human–robot interaction observations, which hardware-oriented roboticists are often unfamiliar with or are not trained for.

Nonterminal failures and workarounds are important for two reasons. First, they directly affect the rate of adoption. A robot mission can be successful but also lead stakeholders to reject the robot as something too difficult, too unreliable, or too limited in functionality to warrant investment of their limited dollars and time. Second, nonterminal failures and workarounds indicate gaps where designs could be improved—a type of "early warning" detection for a feature that will later become critical.

Given the lack of consistent data, a table of nonterminal failures would be difficult to construct; instead, this section provides a compendium of nonterminal failures and workarounds, most of which have been in the rescue robotics literature since 2006 (Murphy, Stover, and Choset, 2005; Murphy and Stover, 2006a; Murphy et al., 2006a; Murphy et al., 2009b; Murphy, 2010b).

There are **five common known nonterminal failures**: *tether tangling, human depth perception, sensors become fouled, "glitches" that cause the robot to be rebooted*, and *breakage*. Tethers tangling has been discussed earlier; in general, a tether requires a separate tether manager to help prevent tethers getting tangled or to get them untangled quickly. Another problem is the frequent repositioning of the robot when the operator could not either gauge depth accurately or precisely control movements, especially when working in vertical orientations where gravity tends to influence motions. Another example of a nonterminal failure is sensors fouling. While this can be a terminal failure, there is the reluctance to remove a robot once it is in place. At the Crandall Canyon Mine disaster, after the robot was removed and its sensors cleaned, they became fouled again when the robot was reinserted. But rather than remove the robot again, the operators continued the mission, got into the mine, and then positioned the robot to use water dripping from the mine ceiling to clean (partially) the mud off the cameras. Robots may experience inexplicable errors and need to be rebooted. And robots break; for example, the Xtreme at the Midas Mine collapse had major damage to its track system, but despite the noticeable loss of mobility, the operators were able to use gravity and tethers to help the robot penetrate to the desired depth.

Workarounds fall into two major categories: those that require **physical or software kludges** and those that require human adaptation. The lack of interoperability or any type of "plug and play" compatibility between robots and sensors is a persistent problem. It was first seen at the World Trade Center in 2001, where a multiRAE gas sensor was duct taped onto a robot to determine if there were explosive gases or enough breathable oxygen in areas where engineers wanted to assess the condition of the basement walls. At Fukushima Daiichi, 10 years later, a recording radiological sensor was zip tied to the T-Hawk UAV in order to get low-altitude surveys. The inability to adapt highly advanced (and expensive) technology to new situations is particularly frustrating as studies in the diffusion of innovations (Rogers, 2003) show that about 50% of the variance in rate of adoption is explained by characteristics of the technology itself; that is, a flexible, easy to adapt robot will be adopted in half the time as a "one-trick pony."

The second category, **problems requiring human adaptation**, is equally challenging. Examples include *poor optics on the robot, poor interfaces that are* *difficult to use with gloves, helmets, and other personal protection equipment,* and *the inability to record (or play back) sensor data or easily to distribute data to remote decision makers.* The operators and stakeholders either "make do" or have to perform multiple extra steps. Certainly, the lack of depth perception and situation awareness is well known, and our studies show that the inability to comprehend complex, destroyed environments affects not just navigation but also searching for victims and understanding dangerous situations (Casper and Murphy, 2003; Burke and Murphy, 2004; Burke et al., 2004; Murphy, 2004a). But there are seemingly minor annoyances that can significantly affect operations. For example, there is an unofficial rule of thumb that any robot with headlights or illumination to work in the dark will have the headlights either at a fixed (wrong) intensity for the conditions or if the intensity of the headlights can be adjusted, the headlight will be placed at an angle not quite where the camera is looking. In terms of expecting stakeholders to absorb more duties, a recent tabletop exercise on the use of UAVs for wilderness wildfires comes to mind (Murphy, 2010b). There, I witnessed several UAV manufacturers try to explain that it would take a few extra steps by someone to move the robot video to a laptop where the data could be converted to a format compatible with the agency's GIS system and that someone could then add labels or flags showing priority (i.e., look here). Not surprisingly, the agencies present replied to the effect that they did not have extra people to do that manually and for the perceived high price of the UAVs, why wasn't that included?

Both nonterminal failures and workarounds are typically handled by the operator or stakeholder having to adapt and work harder to compensate. Human adaptive capacity is not an infinite resource, especially during a disaster where operators are tired, working in uncomfortable surroundings, and are themselves experiencing physiologic stress from the time pressures and emotional impact of the disaster. The inability of the human successfully to overcome design limitations is a strong argument for assistive autonomy.

2.3   General Trends

The general trend is that unmanned systems are increasing in use worldwide, shifting from small UGVs in the United States to a diversity of ground, aerial, and marine assets used in nine countries all over the world. Of the 29 incidents in table 2.1 where robots have been present, UGVs have been at 22, small UAVs at 7, and UMVs at 7 incidents. However, looking at the *total number of robots* deployed versus modality, there have been 34 UGVs, 10 UAVs, and 31 UMVs.

The robots have been used in nine countries, most frequently in the United States and Japan, but also Germany, Italy, Haiti, China, New Zealand, Thailand, and Cyprus.

UGVs have been used at nine underground mines, three building collapses, three earthquakes, one hurricane, one nuclear accident, and one landslide. In seven of the mine disasters, the nuclear accident, and the hurricane, the robots were used in spaces that it was physically possible for a responder could have entered but only at extreme personal risk. In the nine remaining disasters, UGVs were used for extremely small spaces. UGVs have been tasked most frequently for "camera on wheels" search and reconnaissance and mapping missions, but missions requiring manipulation of the environment, such as to open or close doors, vents, and valves or to remove rubble, are becoming more frequent.

UAVs have been used in five earthquakes, three building collapses (accidents, nuclear), two hurricanes, and one flood. Rotor-craft UAVs are deployed more frequently than fixed-wing, possibly because strategic fixed-wing UAVs are available and thus tactical fixed-wings are not needed. UAVs have been used for general reconnaissance and mapping and structural inspection missions.

UMVs have been used for two hurricanes, two earthquakes, two damaged underwater structures (bridge, offshore oil rig), and two recovery (ship, missing person) operations.

ROVs have been used more frequently than UUVs and USVs. Missions for UMVs have been primarily structural inspection of underwater infrastructure and estimation of debris.

2.4   Surprises, Gaps, and Open Research Questions

At this point, a new, perhaps broader, view of rescue robotics is emerging in the reader. However, there remain many misconceptions about disaster robots that merit special attention. The discussion of the performance of the robots highlights an overarching gap in data collection and analysis of disaster robotics, but determining what to measure with what methods and how to extract key insights is an open research question.

2.4.1   Surprises

To reinforce this new view and explode any lingering myths, it may be helpful to examine the most common surprises expressed by students, colleagues, and audiences about rescue robotics. The surprises do not reflect any ordering of frequency of occurrence.

Only one urban search and rescue team in the United States owns a robot. New Jersey Task Force 1, a state team, not a FEMA team, is the only US&R team in the United States known to own and maintain a rescue robot. Federal US&R teams are not allowed to purchase rescue robots because they have not been approved. The Department of Homeland Security, which FEMA is a part of, has been sponsoring the creation of standard test methods (see ASTM E2828) that would verify that a robot model met the minimum qualifications for a rescue robot. No Japanese agency owns a robot, though British fire rescue departments have been actively acquiring all modalities of rescue robots. MSHA owns the V2 robot, but that is specialized for underground mine rescue.

Robots are not used until an average of 6.5 days after a disaster. While robots have become ubiquitous in U.S. military operations, they are still quite new to disaster response. This explains in part the average of 6.5 days between a disaster and when a robot of any modality is used; if the incident command institution had a robot or an existing partnership with a group that had robots, there was an average of 0.5 day before the robot was used, while if not, the average went up to 7.5 days. The socio-organizational culture of response and adoption constraints will be discussed in chapter 6.

There is no single perfect rescue robot for any modality. Both the diversity of missions described earlier and the types of robots deployed shown in table 2.1 offer a strong counterweight to the notion of a perfect robot. I've been asked numerous times by manufacturers and agencies alike what would my perfect robot be? This assumes that there could be a single, Swiss Army knife robot capable of all missions. My reply is, "Are we talking land, sea, or air? If ground, is it a human-scaled space still navigable by humans or a collapse? Is the atmosphere explosive? Water?" And so on . . . until the questioner gets the drift.

UGVs dominate deployments, but UAV and UMV deployments are increasing. While UGVs often receive the most attention, UAV and UMV deployments are increasing. Of the 29 incidents in table 2.1 where robots have been present, UGVs have been at 22 incidents, small UAVs at 7 incidents, and UMVs at 7 incidents. However, looking at the *total number of robots* deployed versus modality, there have been 34 UGVs, 10 UAVs, and 31 UMVs.

Totals can be misleading, as in some cases a robot is used in more than one incident (particularly for mine disasters) or many instances of the same type of robot may be involved (as for the Deepwater Horizon leak), but the numbers do serve to highlight three trends: the increased use of UMVs, the use of multiple modalities of robots at an incident (e.g., Hurricane Katrina,

Fukushima Daiichi), and the use of multiple robots within a modality (e.g., Deepwater Horizon, Pike River Mine).

One possible explanation is that UAVs and UMVs offer a significantly new tactical capability over ground vehicles. Certainly, responders have been dealing with rubble on the ground since the inception of modern disaster response during the World War II Blitz on London, giving 70 years to develop workarounds and to accept a slow pace. But UAV and UMV technology has only emerged within the past decade, so they bring a novel, complementary set of information and capabilities to the responders. Whether this complementary data is more valuable overall than what is provided by ground vehicles is hard to determine; in the end, it may be that the "new and different" of UAVs and UMVs add value and attract attention over the "more of the same" capabilities offered by UGVs. Each modality is discussed in chapters 3–5.

Mine disasters are the most frequent users or requesters of rescue robots. Technologists often make the very human mistake of trying to design the perfect robot for the last well-publicized disaster. However, the numbers suggest that the *opportunity for rescue robots around the world to make a difference is in underground mine disasters*. Out of the 29 disasters where robots have been used or on site, 12 (42%) were underground mine incidents. Of the four disasters where robots were on site but could not be used, 3 (75%) were underground mine events. Furthermore, a request was made for a robot to help with the search for the missing miners at the 2010 San Jose Copper–Gold Mine collapse. Mine disasters are discussed in more detail in chapter 3.

Robots are primarily commercial off-the-shelf technology for civilian applications. Many technologists have a romantic notion of creating the perfect robot on the spot for a disaster. But in practice, and with good reason, rescue robots are not created for a particular disaster; instead, an existing robot(s) with a track record of success is matched with a disaster. All but two robots used in disasters were available for purchase by companies or individuals [known as *commercial off-the-shelf* (COTS)], and the majority of the platforms were already in use for civilian—not just military—applications such as pipeline inspection, aerial photography, ship inspection, and so forth. The two exceptions, the Active Scope Camera and Quince from Japan, came from academia, but those were developed specifically for disaster response and had been tested for close to 2 years by Japanese and American responders in high-fidelity conditions before being deployed. Indeed, every robot but one had been used in the field prior to deployment—and that one, the Mine Cavern Crawler used at the Crandall Canyon Mine, was an assemblage of OEM inspection robot components used previously in dozens of models of waterproof pipeline inspection robots and operated the same way so there was a high confidence that it would work as expected.

Rescue robots typically operate in GPS-denied environments with wireless limitations. It should not be a surprise that building collapses, mine disasters, and interiors of buildings interfere with GPS and any type of radio transmission at any frequency, though it may be surprising to realize how many missions occur in GPS- and wireless-denied environments. The World Trade Center collapse was an extreme example where a primarily steel structure was compacted into dense rubble. But even intact buildings present problems; for example, the thick concrete walls prevented wireless communications inside the reactor buildings at Fukushima Daiichi, requiring UGVs to use fiber-optic tethers (Nagatani et al., 2011). Even being outdoors but operating close to structures may lead to unpredictable GPS or wireless outages in UMVs and UAVs, such as seen in the inspection of the Rollover Pass Bridge after Hurricane Ike (Murphy et al., 2011b). In addition, wireless systems have to function reliably over long distances, as operators may be several kilometers away from the hot zone in a chemical, biological, radiological, nuclear, and explosive (CBRNE) event.

Tethers are essential. Many roboticists look at tethers as the enemy, ignoring that ground robots often have to descend vertically for at least a portion of their run and thus need a belay line and further ignoring the wireless-unfriendly environments these robots must work in. Rather than try to eliminate tethers, many responders have asked for hybrid tethered/ wireless solutions where the robot could move for short distances independently and then reliably reconnect.

The data show that tethers are pervasive. The majority of UGV and UMV (and one UAV deployments) relied on a tether of some sort for power, communications, recovery, or compliance. Many of the smaller UGVs rely on tethers to reduce size and weight while creating an infinite run-time by putting the power supply by the operator. Batteries are a major contributor to the weight of a robot; a longer run-time means a heavier battery even with the advances in battery chemistry. More weight means larger motors, which means heavier robots, which means more power is needed, which means big batteries, and so on. However, even a UGV with onboard power may use a tether, either for communications or safety. An even more complicated scenario is when a robot has two tethers to keep up with; for example, the V2 mine robot is deployed with a fiber-optic cable to ensure communications but often has a second, separate safety line to help lower the robot down the steep slopes in mines. If a robot has a power and communications tether, the tether is strong enough in some, but not all, robots to serve as a belay line.

Only six deployments are known not to involve tethers for at least one of the robots, and all were either aerial or marine. The UAVs at Hurricane Katrina, the L'Aquila earthquake, the Haiti earthquake, and Fukushima Dai-ichi did not have a tether, but a tether was attached to the iSENSYS UAV at the Berkman Plaza II collapse in order to remain in compliance with Federal Aviation Administration (FAA) regulations (Pratt et al., 2008). The USVs at Hurricane Wilma and Hurricane Ike also did not use tethers. All of the deployments involving UGVs appear to have at least one robot that had a tether or belay line. Robots working in the deep interior of the Fukushima Daiichi reactor buildings or beyond the line of sight in buildings used fiber-optic communication tethers, though the ones working on the exterior did not. The UGVs at the Pike River Mine explosion are presumed to have used fiber optics in order to maintain communications at those distances.

While tethers were related to five terminal failures described earlier, they can be equally helpful. Our analysis of the use of tethered robots in the response phase of the World Trade Center collapse showed that while the operator had to manipulate the tether about *once a minute to keep it from becoming tangled*, the operator would also manipulate the tether about *twice a minute to help the robot* drive deeper down into the rubble or navigate over or around obstacles (Murphy, 2004b).

All robots, even ones with autonomy, have been teleoperated. All robots were teleoperated, *even if they were capable of autonomy*. Only four robots had autonomous capabilities, and that was limited to waypoint navigation and did not include any mapping or computer vision assistance. One reason for the lack of autonomy is the lack of sensors: A robot cannot determine how to act without sensing. For ground robots, the small size of the robots used in collapses prohibits the addition of range sensors needed to work in GPS-denied environments, and the lighting (dark) and environmental conditions (dust) are unfavorable for computer vision algorithms that can run on the video imagery. The four robots that had autonomous capabilities, the AEOS Sea-RAI and the three UAVs Raven, Skylark, and T-Hawk, did not use it for at least three reasons. First, for the USV inspecting bridge damage after Hurricane Ike, GPS was neither accurate enough nor reliable enough for operating close to the structure (Murphy et al., 2011b). Second, search, reconnaissance, and many other missions are remote-presence missions, where a human decision maker(s) wants to see and direct data gathering in real time rather than examine data collected by a preplanned route after the fact. The time dependencies and the inability to predict what is valuable to see in further detail or manipulated argue for a human in the loop. Finally, the T-Hawk pilots preferred to maintain positive control of the UAV in case something unexpected happened; by being actively in control, they knew what they were commanding the robot to do, and if the UAV behaved unexpectedly, it was not the software. Note, this is one way of handling the **human out-of-the-loop (OOTL) control problem** frequently seen in automation (Kaber and Endsley, 1997).

The **majority of terminal failures are due to poor human–robot interaction**. That 50% of terminal failures were related to human error should not be surprising given that all the robots were teleoperated. What *is* surprising is that teleoperation is not new (it was an integral part of the space program from before the Apollo program) *and* that many of the robot systems contradict known cognitive engineering and computer interface principles.

2.4.2 Gaps

This chapter highlights a gap in data collection and analysis of disaster robotics. The current state of the practice for reporting deployments is ad hoc, with descriptions appearing in a variety of scientific and trade publications or press releases. There is no guarantee that all deployments have been recorded, much less documented in a manner to support scientific understanding or improved devices and concepts of operations. There is a sparseness of quantitative data about the performance of rescue robots, as many reports are narratives or anecdotal summaries. There is no agreed upon set of data to be formally collected at a disaster to document the use of a robot and no accepted analytical methods. This book attempts to facilitate the move to formal reporting: This chapter reviewed one taxonomy for analyzing robot performance, and chapter 6 recommends a set of data to collect in the field.

2.4.3 Open Research Questions

Subsequent chapters will discuss open research questions by modality, but this chapter puts forward a cross-cutting topic that requires more research: *how to predict robot success for a given mission and set of environmental conditions?* The current state of the art is for the roboticist to work with the responders to make a judgment call, which is captured as heuristics for chapters 3–6. As an aside, I often find myself saying "We used Robot X at these two collapses, thus we believe they will be of great value for your collapse," while wincing that I may be generalizing from too few data points.

There has been some notable work on this topic. Attempting to formalize the choice of a robot for a disaster event was the subject of John Blitch's

1996 master's thesis at the Colorado School of Mines (Blitch, 1996), 4 years before robots would actually be used under his direction at the 2001 World Trade Center collapse. Progress has been made in quantifying robot performance within the DHS/NIST Standard Test Methods for Response Robots used in their Response Robot Evaluation Exercises and are captured in the ASTM E2828 Standard Test Method for Evaluating Emergency Response Robot Capabilities. However, how performance on these single standard tests relates to systems performance in an actual disaster has not been analyzed, in part because larger robots have been evaluated rather than the smaller robots that are more frequently used in disasters. Recent work by Onosato, Yamamota, Kawajiri, and Tanaka (Onosato et al., 2012) has taken a different direction of trying to analyze how buildings collapse in order to determine the appropriate robot for use.

2.5 Summary

Rescue robots are unmanned systems used as *tactical, organic, unmanned systems* that enable responders and other stakeholders *to sense and act at a distance from the site of a disaster or extreme incident.* They are usually COTS devices smaller than military robots and hardened for the extreme environments and must be capable of working without GPS or wireless networks. The typically high cognitive load of working through a robot is exacerbated by the psychological and physiologic stress of being at a disaster. Most of the land and aerial vehicles used in rescue robotics can be traced back to DARPA programs, while marine vehicle development has largely stemmed from the undersea industry.

The economic justification for robots (and associated costs of training and maintenance) is based on four impacts: *saving lives, improving the long-term health and recovery of survivors, mitigation,* and *accelerated economic recovery*. Disasters and extreme incidents occur frequently, and their impact is increasing. Robots can make a difference if they are deployed in a timely fashion, though currently it takes almost a week for a robot to be used.

This chapter answers the four questions posed in the introduction:

Where have disaster robots been used (and for what)? Ground, air, or marine rescue robots have been used for 34 known disasters or extreme incidents starting with the World Trade Center in 2001 and through to the 2011 Tohoku earthquake and tsunami, the 2011 Fukushima Daiichi nuclear accident, and most recently the 2012 Finale Emilia earthquake. The majority of deployments have been to mine disasters or to collapses of urban structures from terrorism, accidents, or meteorological or geological events.

As of April 2013, robots have not directly assisted with saving a life but are credited with speeding the search for survivors, reducing risk to responders, and accelerating economic recovery. Robots are still uncommon, and most agencies or stakeholders do not own robots, preventing the timely incorporation of robots into the response activities.

Rescue robots are generally thought of for the immediate lifesaving response for meteorological, geological, man-made, and mining or mineral disasters. But these robots can be used for the recovery operations as well as prevention and preparation. They can perform immediately and directly engage missions such as search, reconnaissance, mapping, structural inspection, in situ medical assessment and intervention, and direct intervention for mitigation. Robots can also carry out important indirect tasks such as acting as a mobile beacon or repeater, removing rubble, providing logistics support, serving as a surrogate for a team member, and adaptively shoring a collapse to make it safer for responders. Recovery operations include all of the above tasks, but robots can also assist with victim recovery and estimation of debris volume and types.

Often, researchers and investors focus on or are only aware of immediate lifesaving activities in a disaster, neglecting the lower-profile but high-value missions for both rescue and recovery. There is also a tendency to think in terms of replacing a person or a canine, rather than extending and supplementing those capabilities. While incidents like hazardous materials responses could benefit from robots that could do what a human would do if they were able to survive long enough, the majority of missions will use robots for novel tasks. These formative uses of robots pose design risks, as the developer has to project how the robot would fit into larger environmental and organizational constraints.

How well do disaster robots work? Rescue robots work very well, though far from perfectly. While our studies suggest the mean time between failures for rescue robots is fairly low by manufacturing standards, of the order a failure for every 20–100 hours of operation, fortunately missions are much shorter than 20 hours. Robots have only "died in place" in five incidents, a low rate given the extreme environments that they operate in. Fifty percent of the failures are due to human error, which strongly suggests that human–robot interaction is a, if not *the*, major barrier facing rescue robotics.

The history of deployments provides evidence that the current robots are "good enough" to be of value in actual disasters. In the United States, the lack of awareness of robot successes, combined with unfavorable regulations and acquisition procedures, contribute to a dismal record of using robots after they would likely have contributed the most. Disaster robotics is a formative field, with the full range of applications still yet to be discovered, and until they are used more frequently, it will be difficult to create devices, interfaces, and standards that will make rescue robots as common as canine teams.

How can disaster robot performance be formally analyzed? Disaster robot performance is difficult to capture though it is obvious whether or not the robot accomplished the mission, even if it died in place. As there is no way in a real disaster to determine the ground truth, it is not easy to determine if a robot accomplished the mission optimally, was resilient to conditions that it did not encounter, or missed an important cue of a victim or structural hazard. The current state of reporting and analyzing robot performance in the field is ad hoc. This chapter describes a comprehensive failure taxonomy model that requires data as to the environment, the physical subsystems of the robot, and the human–robot interaction. DHS/NIST Standard Test Methods for Response Robots and the resultant ASTM E2828 Standard Test Method for Evaluating Emergency Response Robot Capabilities provide a set of tests of general capabilities, but whether these tests predict actual performance during a mission or are complete has not been determined; the field of disaster robotics is still formative, and thus standard test methods will continue to evolve.

What are the general trends about disaster robotics? While UGVs often receive the most attention, UAV and UMV deployments are increasing. Of the 29 incidents in table 2.1 where robots have been present, UGVs have been at 22, small UAVs at 7, and UMVs at 7 incidents. However, looking at the *total number of robots* deployed versus modality, there have been 34 UGVs, 10 UAVs, and 31 UMVs. Robots are used most frequently for underground mine events, building collapses, earthquakes, and hurricanes. As of April 2013, UGVs have been used successfully at nine underground mines, three building collapses (terrorism, accidents), three earthquakes, one hurricane, one nuclear accident, and one landslide. UAVs have been used in five earthquakes, three building collapses (accidents, nuclear), two hurricanes, and one flood. UMVs have been used for two hurricanes, two earthquakes, two damaged underwater structures (bridge, offshore oil rig), and two recovery (ship, missing person) operations.

What are the surprises, barriers, and research gaps? The primary barrier to deploying robots is not a technical issue but an administrative one. There are multiple stakeholders who can use robots, from federal agencies to municipalities and nongovernmental organizations (NGOs). Only one search and rescue team, New Jersey Task Force 1, and one agency, the MSHA, in the United States own rescue robots. As a result, the majority of robots used at disasters have been provided by CRASAR and CRASAR's Roboticists Without Borders program. Companies and universities who offer or bring robots risk being perceived as profiteers, too high risk, or too experimental if they do not have prior ties with the incident command structure.

Subsequent chapters will discuss the gaps and open research questions for each modality, but the gaps and research questions that cut across all modalities relate to the measurement of performance. The gap between creating new hardware and software for a disaster and being able to measure the performance quantitatively will continue to be a challenge. This trend is expected to continue as more biomimetic platforms (such as snakes, legged, "soft" robots, etc.) and teams of robots are introduced with new failure modes and opportunities for human errors.

Subsequent chapters will discuss open research questions by modality, but this chapter puts forward a cross-cutting topic that requires more research: *how to predict robot success for a given mission and set of environmental conditions.*

3   Unmanned Ground Vehicles

This chapter focuses on the use of unmanned ground vehicles (UGVs) for response and recovery. As seen in chapter 1, UGVs have historically been used more frequently than aerial or marine vehicles. UGVs also are not subject to the policy-based restrictions imposed on unmanned aerial vehicle (UAVs) that will be discussed in chapter 4. UGVs act locally in relatively small areas; individual robots do not cover large areas. While UGVs can be used in either outdoor or interior settings, they have been predominantly used in GPS- and wireless-denied interiors of urban structures such as buildings and mines that have been destroyed by terrorism, meteorological events including earthquakes and hurricanes, or industrial accidents. Whereas multiple robots have been present at disasters, they have not been used as a coherent, coordinated, multiagent system; they have been deployed either individually or in ad hoc groups to support visibility for manipulation tasks.

By the end of this chapter, the reader should be able to:

• Describe the three sizes of land mobile robots: maxi, man-portable, and man-packable.
• Describe the three environments (robot's operational envelope, point of ingress/egress, and operator's environment) involved in selecting a UGV.
• Describe the common missions and tasks for UGVs.
• Match a UGV with the environment and the mission using the mission environment plot.
• Demonstrate familiarity with where UGVs have been used.
• Demonstrate familiarity with the performance of UGVs and their common modes of failures.
• List the advantages of, disadvantages of, and strategies for use of tethers.
• Describe the recommended minimal capabilities and human–robot ratio for a UGV.

• Match a UGV with the environment and the mission using the mission environment plot.
• Demonstrate familiarity with the gaps in the technology and open research questions.

   This chapter begins by describing the three types of UGVs and then focusing on the class of UGVs that has historically been used for search and rescue: mobile robots. The chapter then discusses the environment for a UGV, first broadly discussing application environments and then delving into a formal analysis of the three distinct environments that affect a mobile robot. The robot's operational envelope is composed of one or more regions that may have distinct scale and traversability attributes that will affect mobility and sensing. While the operational envelope is the nominal focus of the robot design, the chapter also describes how a deployable system must also consider how the robot will get in and out of the site (or between regions) and where the operators will be working. With the environmental constraints established, the chapter turns to discussing the missions and tasks for UGVs, noting that there are potentially more missions that UGVs could perform with advances in manipulation. The chapter describes where UGVs have been used, their generally successful performance, and what can be learned from mission failures. The previous experiences suggest a 2:1 or 3:1 human–robot ratio to prevent the currently high incidence of human error. The chapter describes when to use tethers and how to use them. The environments, missions, and history of deployments are brought together in the discussion of selection heuristics, where a novel mission environment plot is used to help guide the choice of UGVs for particular disasters. The chapter concludes with a list of surprises for designers, the observed implementation gaps, and open research questions.

3.1   Types of UGVs

A UGV is a system consisting of: the robot platform, the **operator control unit (OCU)**, and one or more humans operating and collecting data from the UGV. UGV platforms fall into three broad categories: **mobile robots**, **biomimetic robots**, and **motes**. *Mobile robots* is the term for the more traditional mechanical robots, such as bomb squad robots or planetary rovers; the term *mobile robot* originated to distinguish these devices from factory robot arms that were fixed in place. *Biomimetic robots* are those that imitate biology, particularly in terms of locomotion. Humanoid robots such as the Asimo and legged robots such as Big Dog are biomimetic, as are snake-like robots. *Motes* are robots so small that their mobility depends on the wind or other influences in the environment; most roboticists view motes as **unattended ground sensors** rather than UGVs as they are not vehicles explicitly providing mobility. This chapter will use the term *mobile robots* in the sense of field mobile robots, as distinguished from indoor UGVs used for telecommuting or entertainment, and will use the term *mobile robots* interchangeably with the term *UGVs*.

3.1.1 Mobile Robots

Of the three categories of UGV platforms, mobile robots have dominated rescue robotics, as biomimetic robots have been deployed in only two of the 18 UGV deployments. Field mobile robots vary in size but generally use some sort of tread or track for locomotion. The track may be open or exposed (which risks debris dislodging the track) or fixed to the sprocket. Robots used for rescue rely on battery power rather than internal combustion, in part to simplify transportation and shipping.

There are multiple classifications of mobile robots, with no clear consensus. For the purposes of this book, mobile robots will be treated as **robot systems** (robots plus operator control unit and any necessary batteries, fuel, etc.) and will be categorized by their **deployability**: how people get the robots to the point of ingress for the mission. The three basic types of mobile robots, with examples in figure 3.1, are as follows:

• **Man-packable** Robot systems that can be carried in one or two backpacks along with the person's personal protective gear. Note that man-packable means more than being able to strap a robot to a backpack frame; it also means that the center of gravity is such that the person can carry it safely over rubble, up and down ladders, and for long distances. Man-packable robots can be further subdivided into **micro** and **mini**. Microrobots fit completely in one pack and can be deployed in spaces where humans or canines cannot fit. The robots are generally small, tank-like, and the size of a shoebox; the Inuktun Micro-Tracks, VGTV, and VGTV Xtremes are the most commonly deployed version. Note that not all man-packable microrobots use tracks. For example, the TerminatorBot (Voyles and Larson, 2005) starts out as a 3-inch (7.6-centimeter) diameter cylinder, then when it is inserted through a hole cored in concrete, two arms pop out and drag the body along, much like the severed torso of the Terminator pursuing Sarah Connor through the factory in the movie *Terminator*. Some proposed microrobots include ball-like robots, which can be thrown into a building or area and then roll or hop to explore the area.

• **Man-portable** Robot systems that can be lifted or carried by one or two people but do not fit on a backpack or would distort the person's center of gravity. These are also tank-like robots but typically have a manipulator arm. Examples of man-portable robots often used at deployments are the QinetiQ Talon and iRobot Packbot with manipulator arms. These robots work in spaces that a small person might be able to crawl around in.

• **Maxi** Robot systems that are too large or heavy to be manually carried. They generally are transported with a trailer and ramp to the area of interest. Maxirobots may have wheels in addition to tracks and carry manipulator arm(s) and often separate sensor masts that can be raised to provide a more favorable view. Maxi-sized robots are rarely used at disasters, both because of the logistical challenges and because their weight could cause a secondary collapse in a building; however, the V2 variant of the Remotec's ANDROS Wolverine has been involved in six mine rescues or recoveries.

3.1.2 Biomimetic Robots

Biomimetic robots have been used in two deployments and offer promising advances for the future, though most of the research is still focused on controlling the movement of the platform and not the larger set of issues associated with building a robot capable of performing an actual mission. Snakes, or more formally serpentine robots, attempt to duplicate the small size and ability of a snake to move rapidly through difficult terrains. **Free serpentine** robots, which are free-standing and locomote through direct contact with a surface, were used at two structural collapse events. **Fixed-based serpentine** robots, where the snake is attached to another robot like an elephant trunk, have been proposed to allow a traditional robot to use the snake to probe into smaller voids. Robots with legs have become increasingly mature and offer greater mobility, such as stepping over debris, and potential replacement of a human. The DARPA Robotics Challenge is expected to advance humanoid robots, which could allow robots to substitute for humans in extreme situations such as the presence of radiation at the Fukushima Daiichi nuclear incident. Legged robots do not have to look like legs. For example, the rHex robot has rotating curved springs for legs that mimic the motion and adaptability of the joints in insect legs. The slapping motion tends to stir up dust, which can interfere with sensing. Gecko or lizard robots that can adhere to walls have also be proposed for search and rescue, though they have not been shown to work in the dirty, wet, and highly irregular topology typically found in disasters. The *Handbook of Robotics* (Murphy et al., 2008b) has a more complete discussion of promising trends in biomimetic robots.

Figure 3.1
Examples of the three sizes of mobile robots. (a) Maxirobot: an ANDROS F6A. (b) Man-portable: a Foster-Miller Talon. (c) Man-packable: an Inuktun VGTV Xtreme on the left and an Active Scope Camera on the right.

3.2 Environments

The choice of a robot depends on the ecology: how well the robot's capabilities fit the mission needs and the environment. The missions and environments can be very challenging to capture from an ontological perspective. For example, a chemical spill or wildland fire can extend for kilometers, with robots working locally within the area. The same robot might enter a structure and lose GPS and wireless connectivity. Perceptual demands likewise can vary, with some missions being less demanding than others. A robot that is sampling for radiation may not have to accurately record the locations where it samples, while one that is sampling for gases may have to take a physical sample at a precise location and height. A robot that is looking for signs of survivors or signs of imminent structural collapse may be looking for scraps of cloth, droplets of blood, or fine cracks in the structure.

Within the larger global context, *a robot must successfully function locally in three distinct environments*. First, the robot platform must be able to function in the nominal environment (e.g., the rubble). Following standard robotics terminology, this nominal environment will be referred to as the robot's **operational envelope**. The operational envelope is the area of space that the robot is working in or moving through. Second, the robot has to be inserted and recovered; this will be referred to as the **point of ingress or egress**. Third, the robot system has to work in whatever environment the operators are in, called the **operator's environment**. An OCU that doesn't work in bright light is a problem if the operators are likely to be standing outdoors in direct sunlight. Likewise, an OCU that requires the operators to stand within a dangerous area will result in cancellation of a deployment. Although creation of a robot that can navigate and sense in the nominal operational envelope is critical, robot designers often forget the importance of designing for the point of entry or exit and for the human interaction. To be useful, a robot system has to function in all three environments.

3.2.1 Robot's Nominal Operational Envelope

The *operational envelope of a robot reflects its work area*. As a mobile robot, a UGV can move and encounter qualitatively different operational envelopes, which will be called *regions*. Regions reflect patches of the world that elicit a particular set of behaviors. For example, exploring a relatively intact nuclear power plant floor is distinct from climbing up stairs and crossing a narrow catwalk covered with mounds of debris; these would be considered different navigation regions. The example shows how regions can reflect partitioning based on navigational homogeneity, but regions can also reflect partitioning of the operational envelope based on mission needs. Consider that a robot may transit rapidly through one region in order to reach a region where a victim or objective is expected to be.

Formally, a *robot's operational envelope can be divided into a set of one or more regions*. For each region, there are three attributes: the **non-navigational constraints**, the **scale of the region**, and the **traversability of the region**. The non-navigational constraints specify whether a robot can be used for a unique environmental factor; for example, is the robot intrinsically safe for use in explosive atmospheres? The scale of the region determines if the robot can fit and is likely to maneuver successfully in a region; for example, is the robot small enough to fit in the tunnel? The traversability of a region takes into account the conditions within the region that impact navigation; for example, can the robot drive through the mud and over rubble in the tunnel?

3.2.1.1 Non-navigational Constraints Non-navigational constraints fall into four categories: meeting **survivability, sensing,** and **maintainability** requirements and managing **unintended consequences.** Non-navigable constraints have resulted in robots not being used at two disasters (Jim Walter No. 5 Mine and Wangjialing Coal Mine) and experiencing terminal mission failures at two others (Crandall Canyon Mine and Pike River Mine).

Survivability constraints are generally related to the effects of the environment on the robot. They include explosive atmospheres, such as the methane common in mine disasters, high heat from fires, presence of water in the collapse of building structures, and radiation from nuclear events. At the 2001 World Trade Center collapse, robots were expected to navigate through regions where the temperatures were around 177°C, leading to a rubber track falling off (Murphy, 2004b). A robot was prevented from assisting with the Jim Walter No. 5 Mine disaster in 2001 because it was not certified to be explosion proof (Murphy et al., 2009a). Two bomb squad robots not rated for operation in methane-rich atmospheres were lost during the Pike River Mine disaster in 2010 when a second powerful explosion occurred (TVNZ, 2010c), leading to speculation among roboticists that one of the robots set off a spark. Water is almost always involved in disasters, and waterproof robots are now common. If a multistory commercial building collapses, there will be water from the sprinklers as well as from the water fountains and restrooms. Underground mines tap into streams of groundwater. A major concern at the beginning of the Fukushima Daiichi nuclear emergency was how long robots could function in the harsh radiation and would their computer chips, especially the CCD cameras, fail suddenly. Back-of-the-envelope calculations suggested a graceful degradation, later verified by experimentation at a similar Japanese reactor site (Nagatani et al., 2013).

If the robot platform survives and reaches the region of interest, it may not be able to reliably perform the mission due to *sensing constraints*—the environmental impact on *mission sensors* and *wireless network sensors* (i.e., *wireless connectivity*). Lighting conditions and suspended dust, smoke, or vapor particles can interfere with video and range sensors. Video cameras will need external lighting and computer enhancement to see in dark areas, such as deep rubble or mines. External headlights should be tunable, both in the direction of the light and the intensity; invariably, an external light puts too bright a spot on the object of interest. One reason for the lack of sensing reliability may be that military and commercial robots are engineered for specific conditions versus the environments seen at disasters. For example, the headlights on the Solem robot used on World Trade Center Building 4 were too bright for the twisty voids in the rubble as shown in figure 3.2; the furthest straight-line distance was of the order 1.2–2.4 meters while the robot illumination was intended for distances of 6 meters or more (Micire, 2002; Murphy, 2004b). An example from the opposite end of the spectrum, the headlights on the Inuktun VGTV Xtreme used at the Midas Gold Mine collapse in 2007, were designed to illuminate about 3–4.5 meters ahead of the robot, whereas the robot needed to see 20–40 feet ahead; in that case, the robot cameras were pointed at the spot where the victim was lying, but the body could not be detected with the low light (Murphy and Shoureshi, 2008; Murphy et al., 2009a). Video camera streams also need to have exposure gain control as they may also encounter bright shafts of light from above. Range cameras, especially those using infrared sensors, are susceptible to ambient light. Ongoing work by our group has shown that the low-cost RGB-D sensors such as the Kinect that have become extremely popular with roboticists will not work reliably in outdoor disaster conditions (Craighead, Day, and Murphy, 2006).

The reliability of wireless connectivity is difficult to predict in structures and subterranean environments, thus wireless robots have been used at only

Figure 3.2
View from Solem at World Trade Center Building 4. Note the highlights on the left, interfering with gain control to see the path in the center of the image.

two deployments for initial response and recovery, both under very limited conditions. Wireless robots were used at Fukushima Daiichi but only within line of sight, so a wireless robot could do down a corridor and turn to look in a room but not enter (Kawatsuma, Fukushima, and Okada, 2012). At the World Trade Center deployment, the only wireless robot used had a loss of 25% of the video frames (Murphy, 2004b). In practice, a wireless robot is fitted with a fiber-optic cable to ensure communications or is restricted to work in direct line of sight of the OCU. Methods for increasing reliability exist but have not been adopted by military and commercial robot manufacturers. MSHA experimented with a robot dropping off wireless repeaters (see figure 3.3) as needed, like bread crumbs. The DARPA LANdroids project demonstrated miniature robot repeaters that could move to maintain connectivity (McClure, Corbett, and Gage, 2009). Even if connectivity is lost, autonomous return-home capabilities have been commonplace since the 1990s on UGVs with onboard control.

Another practical constraint imposed by the environment is the *maintainability of the robot*. One of the first questions asked by responders about a robot is *can it be decontaminated and brought out of the hot zone, either for recharging, repairs, or to be used elsewhere?* Decontamination most likely summons to mind the Fukushima Daiichi ground and aerial robots, which were coated with radioactive dust. But decontamination is essential for all rescue robots, as these robots will be exposed to debris that may contain asbestos, deadly chemicals, or biohazards such as sewage or even blood and tissue. Cement dust and water produce a corrosive mix that will eventually damage a robot.

Robots that cannot be satisfactorily cleaned in the hot zone have to be left there. Leaving robots in the hot zone may not be a problem if they are going to be extensively used, but it means that any maintenance, repairs, changing of batteries, and so forth, have to be done by an operator in full personal protective gear. But a person in gloves is often unable to reach into the small clearances and handle the types of tools needed to open up the robot and get to batteries or key components. There is also a cultural gap between the robotics and fire rescue communities; waterproof may be interpreted as being able to stay waterproof under the blast of a high-pressure fire hose, the default strategy for cleaning of equipment.

All disasters pose the possibility of creating *unintended consequences*, and these unintended consequences can lead to a robot being discarded. Maxisized bomb squad robots were on site at the 1995 Oklahoma City bombing but were not used because of fears that their weight would cause a secondary collapse and crush survivors below (Manzi, Powers, and Zetterlund,

Figure 3.3
An iRobot Packbot deploying SPAWAR wireless repeaters at an MSHA test at the Jim
Walter No. 4 Mine.

2002). In the Midas Gold Mine collapse, a man-portable Allen-Vanguard
bomb squad robot was lowered into a multistory vertical void within the
mine (Murphy et al., 2009a). The void was irregular, with many ledges that
the missing miner might be lying on. However, the weight of the robot
was such that it was knocking the ledges down, possibly burying the body.
Thus, the robot was removed, and later a man-packable microrobot was
used to search the voids. Some unintended consequences do not eliminate
the use of robots, though their use may not be totally desirable.

An example of the possibility of unintended consequences occurred in
the case of our deployment to the Cologne Historical Archives building collapse in 2009 (Linder et al., 2010). There the responders asked if the robot's
tracks would damage documents—a reasonable question given that some of
the documents buried in the rubble dated to AD 800. Our response was that
the tracks would probably crumble some types of paper and parchment but
would definitely leave a trail of mud on them. While this was clearly not
optimal, the decision was made to use the robot in order to speed up the
structural inspection and location of the most valuable documents, though
the use of the robot was canceled due to a shift in operational priorities.

3.2.1.2 Scale of a Region   The scale of a region reflects the ratio of the overall **characteristic dimension(s) of the environment**, E_{cd}, to the **characteristic dimension(s) of the agent**, A_{cd}, either a person or robot, moving through the environment. For the purpose of rescue robotics, the scale of a region can be classified as either *exterior, habitable, restricted maneuverability,* or *granular* (figure 3.4). Understanding the scale of a region is useful in selecting the size of robot and projecting what regions are suitable for a particular size of robot.

In **exterior spaces,** the environment is much larger than the agent. Interior spaces can be very large compared to the agent or smaller than the agent. Mathematically, the ratio is $E_{cd} \gg A_{cd}$.

One size of interior space is *habitable space* (space either built or naturally occurring), which allows the agent to move easily around without stooping. An example of a built habitable space is a building such as the Fukushima Daiichi reactor units or a structure such as the Sago Mine. Naturally occurring spaces such as caves and caverns can serve as habitable spaces. Here, the ratio is that of an agent having at least two body widths to turn around in, or $E_{cd} > 2A_{cd}$.

The next interior space is *restricted maneuverability space*, where the agent cannot easily move around. This may be due to a narrow or confined built structure where an agent has to crawl or squeeze through, such as tank cars, utility tunnels, air ducts, and so forth. Some mines require miners to stoop. Natural structures such as caves and voids reduce mobility. Here, the ratio is $E_{cd} < 2A_{cd}$.

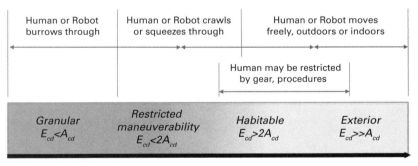

Figure 3.4
Spectrum of scale of regions relative to the physical agent. Light area is favorable E_{cd} to A_{cd} ratios, whereas dark area is unfavorable.

Interior spaces may be smaller than the agent, called *granular space*, and the agent has to burrow in order to move through it. Currently, neither people nor robots work in these spaces, though animals such as the sand-fish do, and interesting biomimetic research is progressing. The ratio for granular space is $E_{cd} < A_{cd}$.

These four partitions are fuzzy. Consider that habitable space may become restricted maneuverability space if a person wears level A personal protective equipment and cannot turn as easily or would like to have more room to make sure the suit does not snag and tear. Some spaces call for restricted maneuverability; for example, if a person is going to enter an underground ("below grade") structure where there is only one way in and out ("confined"), there is usually a set of safety procedures and restrictions. In some cases, the person may need to wear a safety harness with a belay plus carry an air tank, effectively restricting mobility.

The *scale of the region provides insights into the appropriate size and function-ality of a rescue robot.* A reduced maneuverability space for a person might be a habitable space for a man-packable microrobot; indeed, mobile robots have been used for 6 of 18 deployments to go in places people and canines could not physically fit. However, a man-packable microrobot might not be effective in exterior spaces where a person even in protective gear could cover more ground faster.

Figure 3.4 suggests that regions falling in certain portions of the spec-trum favor humans, and thus to have a favorable cost–benefit ratio, robots must perform on par with the human. The exception would be if the envi-ronmental constraints such as safety prevented human entry. This was the case at the forensic examination of the portion of the parking garage still standing after the Berkman Plaza II collapse in 2007, where a robot was used to search a void that a person could have crawled in but where there was a high risk of structural collapse. However, if there had been clear signs of a living person trapped in that space, it is likely that a responder would have risked his life to get to the person, as the robot would have been much slower.

3.2.1.3 Traversability of a Region   The scale of a region is analogous to a geographical region on a map, and as the famous saying goes, the map is not the terrain. The terrain of region consists of the factors that impact the traversability of the region—whether the robot can actually move through the environment even if it can fit. Traversability is a function of the region's **tortuosity, verticality, surface properties, severity of obstacles,** and **accessibility elements.** Figure 3.5 shows how these attributes impact traversability, especially as the terrain becomes more three-dimensional or vertical as opposed to moving in a roughly horizontal plane.

Tortuosity and *verticality* are similar to terrain topology and reflect how much effort it takes to move about, much like following straight paths versus windy roads and driving across a plain versus climbing a mountain. *Tortuosity* refers to the number of twists or turns in a space. *Verticality* is the slope of the terrain being traversed. In structures built for human habitation, verticality is limited to stairs or ramps, but in disasters, humans or robots may need to enter a building from the roof and progress downward to find survivors. Void spaces through collapsed buildings and mines are not only small and thus restrict maneuverability but also include vertical drops and many twists. The more torturous and vertical the environment, the shorter the effective sensing distances need to be; the robot has limited area in direct sensing line of sight before the corridor turns and is hidden.

Surface properties and *obstacles* are similar to vegetation on the terrain and thus can impede navigational progress and sensing as well. *Surface properties* include mud, carpeting, and slippery, wet metal treads on staircases. Surface properties also affect sensing, as light-based range sensors see through glass; mirrors, metal surfaces, and multiple edges reflect and deflect active sensing; and cloth and carpet absorb light and sound signals.

Mud and very long, old-style shag carpeting were two impediments to the robot we used at the La Conchita mudslides. The Inuktun Xtreme

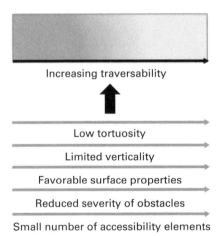

Figure 3.5
Gradient plot of how traversability increases with decreasing terrain challenges.

attempted to navigate fairly horizontal terrains (Murphy and Stover, 2008). In one case, the robot tracks stuck in the mud and came off as the robot moved from the exterior of a house to an area underneath the house. Later, the robot tracks came off again when the robot was lowered through a window into an attic loft to search another house. The tracks had robustly stayed on the robot for concrete, gravel, and pavement, but had not been tested for these other conditions.

In addition to surface properties, the *severity (the number and size) of obstacles* can be an impediment. The more obstacles or clutter, the more the robot has to avoid or negotiate. Obstacles can be objects protruding into the nominal free space of a region, such as chunks of debris in a hallway or on a catwalk; these protrusions are called **positive obstacles** because they are above the plane of travel. Obstacles can be **negative obstacles** if they extend below the plane of travel; these are more commonly referred to as "holes" or slopes or stairs heading down. Positive obstacles are often irregular, where debris may have reinforcement bar ("rebar") or narrow slivers of wood sticking out, which are hard to see. The robot must either stay away from the obstacle in order to avoid becoming snagged or move slowly and carefully.

If tortuosity and verticality can be thought of as the topology of a region and if the surface properties and severity of obstacles can be thought of as the vegetation and overgrowth in the region, then the *number and type of accessibility elements* can be likened to projecting how a person with disabilities can get around. *Accessibility elements* is a term taken from architecture that refers to doors, stairs, ladders, or any component of the habitable space that controls the choice of movement. Accessibility elements are the bane of mobile robots, with doors and stairs requiring painstaking teleoperation or fragile autonomy. The iRobot Packbots at Fukushima Daiichi were unable to climb the metal stairs in the Fukushima Daiichi reactor units to get to the third floor and inspect the unspent fuel rod pools. Consistent with bomb squad experiences, two robots were needed to open doors, one to provide a favorable camera view for the operator of the other robot so that he could quickly manipulate the handle.

Taken together, it is straightforward to see how increasing tortuosity, verticality, unfavorable surface properties, severity of obstacles, and number of accessibility elements drive traversability down. Environments that are easy to traverse such as horizontal, flat surfaces with few obstacles or doors fall on the right portion of the plot in figure 3.5, while irregular vertical voids with debris and ledges fall on the left side of the plot signifying their difficulty. *While all five factors contribute to the difficulty or ease of traversability, in general they group into mostly two-dimensional horizontal regions or a mostly three-dimensional region requiring tortuous vertical excursions.*

3.2.1.4   Sets of Regions   Being able to traverse *a* region is not sufficient: A robot has to be able to traverse *all* regions in the area of interest to be considered successful. *Eight of the 13 UGV deployments encountered at least two distinct regions.* Consider that a robot may be able to move around on one story or floor of a building but be unable to climb stairs to reach the next floor, as happened at Fukushima Daiichi. Perhaps the most striking example is the use of a polymorphic robot, the Inuktun Mine Cavern Crawler, at the Crandall Canyon Mine disaster. Figure 3.6 diagrams how the robot had to be lowered (and crawl at places) through a tight, uncased borehole going straight down through rock, then go through an irregular hole cut in the wire mesh supporting the mine ceiling, and finally transform itself into a robot capable of moving on the horizontal mine floor.

3.2.2   Point of Ingress/Egress
A robot must first get to the regions of interest in order to be functional in the desired regions. The entry, or *ingress*, into the workspace is usually quite different from the regions in the workspace and often requires manual insertion. The point of ingress may be a **natural void** formed in a collapse, an **engineered breach or borehole** through a roof, masonry wall, or debris, or an **accessibility element** on the surface, such as a door or portal. Figure 3.7 shows examples of typical natural voids. The exit, or *egress*, is usually the point of ingress, as most robots have used tethers or fiber-optic cables and must return the way they came or need to stay in the line of sight of the wireless transmitter on the OCU. In all known deployments, mobile robots have been manually carried or joy-sticked to the point of ingress. This may change in the future, as the DARPA Robotics Challenge is exploring scenarios where a robot drives itself to the site of interest and enters the interior.

My original foray into rescue robotics in 1995 examined marsupial robots for ingress, where an agile man-portable robot "mother" would carry a man-packable "daughter" or "joey" robot that would crawl out and enter the void in the rubble (Murphy, 2000). This marsupial concept proved unworkable at the World Trade Center collapse, as the mother robot would have had to crawl to a crevasse, descend nearly straight down for 10 or more meters, then climb back another 10 meters over highly irregular rubble (see figure 3.8). For that response, the robots were carried in backpacks over rubble or down straight ladders.

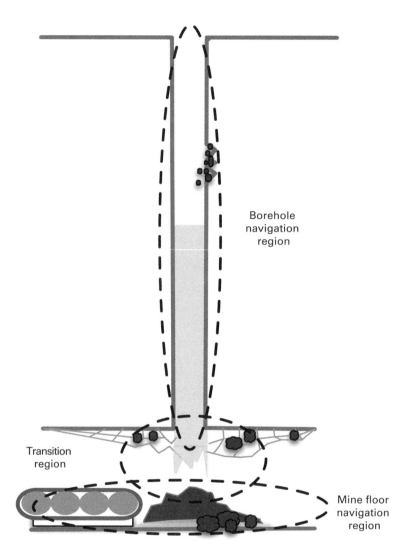

Figure 3.6
Diagram of the set of three regions that an Inuktun Mine Cavern Crawler robot had
to go through at the Crandall Canyon Mine disaster. Light shading is water, black is
rocks and mud, and dark shading is mining equipment.

Figure 3.7
Rubble at the Cologne Historical Archives building collapse. Voids deeper than could be examined by a search camera are circled.

The first type of entry into rubble is through a *naturally occurring void*. Voids that extend into compressed rubble are rare; they are often shallow pockets rather than tunnels. Building materials and furniture often are pulverized, filling any gaps with dirt, and fabrics and carpet deform and fill spaces. The void itself is likely to be irregular in cross section and of the order 1 meter or less in diameter. The voids are often tortuous as well, and a common trick is to insert a pipe into a void to help get the robot into the interior. One exception to tortuosity is the voids found in a pancake collapse of built structures such as seen at Berkman Plaza II, Cologne Historical Archives building, and Prospect Towers collapses. Because pancake collapses are often stacked in layers, robots can follow the layer—if they are small enough.

The second type of entry into an interior space is by a *borehole or breach*. In underground mine disasters, vertical boreholes will be drilled through rock to attempt to intersect the portion of the mine where there may be survivors and to mitigate the atmosphere. A robot was used at the Crandall Canyon Mine disaster to enter two boreholes drilled more than 450 meters deep and search for survivors (Murphy et al., 2009a). Horizontal, and sometimes vertical, breaches are made into masonry structures to provide access. These breaches are made with coring tools that cut a series of holes, usually 3 inches (7.6 centimeters) in diameter, until a triangular space sufficient for entry by a human can be carved out. Coring takes several minutes, so there is a real need for robots such as the TerminatorBot and serpentine robots that can enter a single-cored hole.

Entry through the third type of entry, a *surface accessibility element* such as a door or portal or other intact element designed for human navigation, might appear to be the least demanding. However, this isn't always the case. At the Excel No. 3 Mine fire, the robot had to enter the mine through the surface entry used by the miners, a ramp with a 15-degree slope (Murphy and Shoureshi, 2008). However, the ramp was icy from the use of liquid nitrogen to suppress the fire, and a safety rope had to be attached to the robot. The safety rope was a distraction, and the operator ran over the fiber-optic communications cable while focusing on not tangling the safety rope.

3.2.3 Operator's Environment

The third environmental consideration for a robot system is the *operator's environment*. All mobile robots have an OCU with operators or responders viewing and directing the robot. The environment can dictate where the OCU is placed and even if the OCU, and thus the robot, can be used.

A major consideration in the placement of the OCU and its stand-off distance from the point of ingress/egress is *safety*. For an unstable building collapse or a hazardous material (hazmat) incident, the OCU may have to be placed outside of the collapse or hazmat zone, which can be 0.8 kilometer or greater. This means that even though the workspace for the robot is in the interior of the collapse, the OCU and operator may be outside. The OCU and operator may be exposed to bright sun, hot or cold temperatures, or rain and snow; these all affect the visibility of the screen and overheating of the OCU. If the OCU cannot be placed in a safe location, then the robot will not be used. This occurred at the Cologne Municipal Archive building collapse where the Active Scope Camera robot was the right size for the voids, but to use it required the operator to stand at the point of ingress (Linder et al., 2010). In this case, the point of ingress was below an unsafe slab of concrete where a person could not be permitted to stand. Therefore, the robot was not used. At the Midas Gold Mine collapse, our team waited as a cantilevered deck was constructed next to the void to permit robot operations (Murphy and Shoureshi, 2008).

In addition to safety, another consideration is the *available space* to set up the OCU and for the operators to work in. The available area varies. At the World Trade Center collapse, the CRASAR teams were often perched on steel beams, but at the Midas Gold Mine there was enough room for a table and a large external monitor on the deck.

3.2.4 Case Study: The Impact of the Environment on UGVs at Hurricanes Charley and Katrina

Hurricane Charley (Murphy, Stover, and Choset, 2005) and Hurricane Katrina (Micire, 2002) are examples of why mobile robots may not add value to geographically extensive events such as hurricanes or earthquakes. The environment and expected tempo of operation for search missions favors canines who can detect the presence of survivors through closed doors and over significant distances. The two hurricane deployments illustrate how the overall environment affects the suitability of a robot based on logistics, the ability to function in the operational envelope, the ability to enter and exit, and the location of the operator and OCU. The deployments suggest that mobile robots are better suited for a mule-like follower mission to carry gear for responders than for conducting search, rescue, and victim recovery missions in a habitable space populated with accessibility elements and debris. Figure 3.8 shows what an affected house and surroundings looked like after a storm surge; the photograph was taken after Hurricane Katrina, though not from the building searched in (Micire, 2008).

3.2.4.1 Impact of Point of Ingress: UGV Not Used at Hurricane Charley

The nonuse of robots at Hurricane Charley in 2004 illustrates the limitations imposed in practice by logistics and by accessibility elements for points of ingress and the robot's operational envelope. Hurricane Charley was a

Figure 3.8
House at Bay St. Louis destroyed by Hurricane Katrina. (Photograph taken by Mike Lotre for CRASAR.)

category 4 storm that swerved from its predicted path and hit an area of Florida that had not had mandatory evacuation. CRASAR was embedded with Florida Task Force 3 for our first deployment after September 11. Florida Task Force 3 was the first response team to enter Punta Gorda, which was ground zero for the hurricane and has a large retirement population. Even though it was midnight, squads were dispatched to go door to door to see if there was anyone who needed help. Multistory commercial buildings generally withstand hurricanes and are not occupied during a major weather event, so the immediate focus was finding and assisting people who may not have had time to evacuate. The squad I was on walked about 2 kilometers to the first set of condominiums and apartments right at the point of the peninsula; note that this meant carrying a man-packable Inuktun Micro-VGTV, including batteries for 12 hours of operation. The residences and apartment buildings were largely intact with obvious superficial damage to railings and roofs but no signs of imminent structural collapse. Going door to door presented a conundrum: If no one answered the door or responded to shouts, did you need to check inside to see if a person was either unconscious or hiding/trapped in a closet and you couldn't hear the person's call for help? If you did need to check, you would have to either break down the door or break and enter through a window. This could result in unnecessary damage and leave the apartment vulnerable to looting.

The UGV we had brought—indeed any UGV—would not have been of value in solving this conundrum because UGVs are ill-suited for the operational envelope and point of ingress/egress for an intact residence. First, in order to use the robot, the responders would still have needed to break down a door or a window. Second, the robot was less efficient and capable than a person. A responder could enter, rapidly look around and open doors, whereas a robot in general would be much slower, and the robot we had didn't have a manipulator for opening doors or for sliding screens. Third, which was related to the lack of efficiency for searching a single dwelling, a robot could not search a neighborhood of more than 100 dwellings in a reasonable amount of time. Instead, Charley showed why canines have become an important component of a formal response: They are mobile, fast, and can work at stand-off distances without entering or breaking down doors.

3.2.4.2 Impact of All Three Environments: UGV Used at Hurricane Katrina The use of a UGV at Hurricane Katrina illustrates the impact of the environment on a robot's operational envelope, point of ingress/

egress, and the operator's environment. A man-packable, tethered Inuktun VGTV Xtreme robot was used by a former student of mine, Mark Micire, with Florida Task Force 3 to search two apartment buildings in Biloxi, Mississippi (Micire, 2008). Only data from the search of the first apartment building were reported, but it is sufficient to illustrate the impact of the three types of environments. The first building, a two-story apartment complex, was unstable, leaning to one side with the lower floor reduced by the flood surge to mostly studs and framing, cluttered with a washed-up car and debris. As deceased victims had been found in the other part of the complex, the team wanted to examine the building but could not enter due to the imminent collapse. The robot was too small to climb stairs and had no way of opening doors so it could not conduct the search. To determine how to shore the building to allow the responders to search the upper floor, the robot was sent in for a structural inspection mission to look at interior structural elements on the first floor. The robot's operational environment posed a non-navigational constraint: an audible natural gas leak. As the VGTV Xtreme is not intrinsically safe, the team crimped the gas line to stop the leak and reduce the threat of explosion. The region of interest was human-habitable in scale but with a high severity of obstacles; even though a person could have stepped over the mattress on the floor or the debris, the obstacles were large relative to the robot and posed tipping hazards. In terms of the point of ingress/egress, two entry points near structural points of interest were identified, as trying to explore all of the ground floor from one entry point risked tangling the tether. In terms of the operator's environment, the OCU was set up outdoors near the ingress point on the first run but had to be moved farther back for the second ingress point to stay out of the collapse zone.

3.3   Missions and Tasks

The 18 deployments of UGVs involved seven of the rescue robot missions identified in chapter 1, and six other types of mission were proposed. The most common mission has been to perform reconnaissance and mapping (13 of the 18 deployments). However, reconnaissance and mapping has been used in conjunction with search or victim recovery 11 times, often with an implicit mission of structural inspection to project how to extract a survivor if encountered as noted in Murphy (2004a). Concurrent missions have ramifications for human–robot interaction and present opportunities for autonomy as will be discussed in section 3.4. Understanding the seven missions leads to understanding the types of tasks that a UGV should be competent at. The types of tasks or competencies fall into two broad categories: navigational tasks and mission tasks.

3.3.1 Missions

The seven missions that UGVs have been used for in order of frequency are as follows:

• *Reconnaissance and mapping (13 times)*. Reconnaissance and mapping missions fall into three categories: *understanding the physical state of the area of interest, environmental sampling,* or *both*. In the case of a response to or a remediation of a collapse or explosion, a priority is to understand the structural integrity of the area of interest. However, in a hazardous material situation such at Fukushima Daiichi or the aftermath of a mine fire (e.g., Barrick Gold Dee Mine, Excel No. 3 Mine, DR No. 1 Mine, McClane Canyon Mine) where the structure is likely intact, structural understanding is less critical. Instead, the concern is that explosive gases or lack of oxygen might be present and endanger response or recovery efforts.

• *Search* or *victim recovery (9 times)*. During the initial response phase, robots have been used to search for survivors or later to recover victims. The distinction between searching for survivors or victims is a matter of semantics, as authorities are often reluctant officially to give up hope for finding someone alive.

• *Structural inspection (5 times)*. The structural inspection mission is distinct from reconnaissance and mapping based on the specialized information needed by subject matter experts. While an expert can often get an overall understanding of the state of the physical environment based on reconnaissance, structural decisions often require much more information. In particular, experts need accurate measurements of the lengths and depths of cracks, size of exposed aggregate in concrete, and so forth. A photograph or video that looks good to a nonexpert is generally not sufficient, which is why my research centers on how to incorporate subject-matter experts into the loop without knowing, or driving, the robot.

• *Direct intervention (5 times)*. The robot may need to interact with the environment, beyond opening up accessibility, to mitigate the disaster or repair and restore critical functionality. Robots have been used to assist with the reopening of four mines by pushing aside rubble, opening ventilation doors, readjusting ceiling beams, and pulling out timbers (Murphy and Shoureshi, 2008). The robots at the Fukushima Daiichi nuclear accident also had to intervene. The DARPA Robotics Challenge proposes a direct intervention mission.

• *Rubble removal and clearing (1 time)*. At Fukushima Daiichi, construction robots cleared debris and radioactive materials (Kawatsuma, Fukushima, and Okada, 2012).

• *Forensics (1 time)*. Understanding the cause can be important for legal reasons, such as the liability in the collapse of the parking garages at the Berkman Plaza II and the Prospect Towers, but also for engineers to reconstruct the disaster and learn from it. The robots were used at Berkman Plaza II initially at the request of the Jacksonville Fire Department for the immediate response in trying to locate a missing worker, then a few weeks later to assist the lead structural engineering firm, Bracken Engineering, in collecting unambiguous data about the collapse and the integrity of the standing portion of the garage.

Other missions for UGVs have been proposed or experimented with. These are as follows:

• *Act as a repeater*. The idea of using robots as wireless network repeaters is not new: the DARPA LANdroids program (McClure, Corbett, and Gage, 2009) is one example of previous research initiatives. We helped test and evaluate the SPAWAR repeater system for a iRobot Packbot at the Jim Walter No. 4 Mine (Murphy and Shoureshi, 2008). Despite the large volume of research and experimentation, mobile robots as repeaters have not yet been deployed.

• *Act as a surrogate*. As we discovered during an exercise with Virginia Task Force 2, a small robot with two-way audio was spontaneously adopted by both a responder breaching a wall several meters in a narrow tunnel and the external team members as a way of communicating (Fincannon et al., 2004). The responder could point to questionable areas, while the outside team members could look around and see what the state of progress was and prepare for the next steps, speeding overall efficiency. Responders at a later NASA Ames DART exercise likewise used the robot as a surrogate for team members outside the area of interest.

• *Adaptive shoring*. Our work has shown that a set of robots, more like mobile airbags, can help maintain stability in a rubble pile as debris is removed (Murphy et al., 2006b), thereby reducing the risk of a secondary collapse. This could greatly speed extrication, which must proceed slowly for fear of making matters worse.

• *Logistics support*. One of the recommendations from the experience of not using a robot at Hurricane Charley was to create robot mules that could carry gear back and forth (Murphy, Stover, and Choset, 2005). Our work with the Texas Engineering Extension Service and a focus group of wildland firefighters produced a similar recommendation for wildfires (Murphy et al., 2009b). Logistical support for responders is critical, yet at the same time it is important to minimize the number of people in the hot zone; robots that can help response and recovery professionals to focus on their specialties could be of great benefit.

• *In situ medical assistance.* To date, no robot has made a "live save," but a robot might have to be a surrogate for medical personnel for 4–10 hours until extrication teams could get close to the survivor (Murphy, Riddle, and Rasmussen, 2004). Our work has begun examining this possibility, working with medical personnel to determine concepts of operations (Riddle, Murphy, and Burke, 2005; Gage, Murphy, and Minten, 2013) and triage protocols (Chang and Murphy, 2007) for telemedicine. We have created a provably comforting robot head to be a multimedia "survivor buddy" to support telepresence and to allow the survivor to play music, watch videos, and engage the outside world (Murphy et al., 2011a).

• *Victim extrication and evacuation.* Removing survivors, particularly from a hazardous material event where there may be many people immobilized or disoriented, is another area that is getting increasing attention. The U.S. Army Telemedicine and Advanced Technology Research Center has been leading initiatives in robots such as the Vecna BEAR for autonomous casualty extraction.

3.3.2 Tasks

The missions listed above suggest that a competent UGV will be capable of performing nominal navigation tasks and more sophisticated mission tasks. The nominal navigation tasks are the ability to:

• Move through low-traversability regions, with the capabilities of obstacle avoidance, obstacle negotiation (i.e., climbing over rather than going around), and self-righting or being able to function, including sensing, while inverted.
• Manipulate accessibility elements.
• Manipulate obstacles to get them out of the way.
• Localize position and produce a map, even in GPS-denied environments.
• Return home after losing signal if a wireless robot.
• Monitor/manage vehicle health.

The nominal mission tasks are the ability to:

• *Sample environmental properties*, which encourages "plug and play" mechanisms to allow responders to add their preferred handheld gas or radiological sensor of choice.

• *Support cooperative perception and sensor fusion* to reduce the workload of the operator and experts. This includes image enhancement, cueing the operator to images or readings that may require a further look.

• *Guarantee sensor coverage* of a volume of space or of an object, such as an injured person. This implies that the sensors move independently of the robot navigational platform so that if a robot has restricted mobility, the sensing is still complete.

• *Conduct investigatory manipulation* to probe an object or terrain, providing clues as to whether a substance is hard or soft, a shadow or a liquid, and so forth.

• *Use manipulation to intervene* in the disaster, such as by dropping off sensors, turning valves, or remediating the environment by moving rubble or timbers.

3.4   Where UGVs Have Been Used

Table 3.1 summarizes the 23 events where UGVs have been used (18 events) or were on-site ready to be used (5 events). Each row lists the robots used (if known) for each event by model, type, and mission. The five deployments where the robots were not used are shaded in gray.

The table shows that UGVs have been used for mine disasters (9 times), structural collapses (8 times), and hazardous materials events (1 time). The causes of the structural collapses were accidents (2 times), earthquakes (3 times), terrorism (1 time), hurricane (1 time), and mudslide (1 time). They were not used at the Jim Walter No. 5 Mine because the robot was not mine permissible; for Hurricane Charley because of point of ingress; for the Cologne Historical Archives building collapse because one robot was too large for the voids and the other robot posed operator safety risks; the Wangjialing Coal Mine presumably because the robot was not waterproof; and the Upper Big Branch Mine because that robot was not considered agile or reliable enough.

The table shows that maxi-sized robots have been used most frequently, followed by man-packable mini-sized robots, and finally man-portable micro-sized robots. This seems counterintuitive, as it indicates that the largest and smallest robots are the most valuable. The data on the maxi robots is skewed by the use of robots for mine rescue and at Fukushima Dai-ichi and the limitations of current technology. The mine data are skewed because the only mine-permissible robot in the work is a maxirobot. However, the MSHA and National Institute for Occupational Safety and Health

Table 3.1
Table of known UGV deployments

Year	Event	Robot	Robot Type			Missions							
			Maxi	Mini	Micro	Search	Reconnaissance and Mapping	Rubble Removal	Structural Inspection	Victim Recovery	Direct Intervention	Forensics	Demonstration
2001	World Trade Center (USA)	Inuktun Versatrax	X			X	X						
		Inuktun VGTV	X			X	X						
		Foster-Miller Solem		X		X	X						
		Foster-Miller Talon		X		X	X						
		iRobot Packbot											X
		SPAWAR UrBot											X
2001	Jim Walter No. 5 Mine (USA)	ANDROS Wolverine	X			X							
2002	Barrick Gold Dee Mine (USA)	Mine-permissible ANDROS Wolverine	X				X						
2004	Brown's Fork Mine (USA)	Mine-permissible ANDROS Wolverine	X								X		
2004	Niigata Chuetsu earthquake (Japan)	Souryu			X								X

Note: Shading indicates a robot was on site but not used.

Table 3.1 (continued)

Year	Event	Robot	Maxi	Mini	Micro	Search	Reconnaissance and Mapping	Rubble Removal	Structural Inspection	Victim Recovery	Direct Intervention	Forensics	Demonstration
2004	Hurricane Charley (USA)	Inuktun VGTV			X								
2004	Excel No. 3 Mine (USA)	Mine-permissible ANDROS Wolverine	X				X				X		
2005	DR No. 1 Mine (USA)	Mine-permissible ANDROS Wolverine	X				X				X		
2005	McClane Canyon Mine	Mine-permissible ANDROS Wolverine	X				X				X		
2005	La Conchita mudslides (USA)	Inuktun VGTV Xtreme			X	X	X			X			
2005	Hurricane Katrina (USA)	Inuktun VGTV Xtreme			X	X	X		X	X			
2006	Sago Mine (USA)	Mine-permissible ANDROS Wolverine	X			X	X		X				
2007	Midas Gold Mine (USA)	Allen-Vanguard Inuktun VGTV Xtreme		X	X	X X				X			

Note: Shading indicates a robot was on site but not used.

Table 3.1 (continued)

Year	Event	Robot	Robot Type			Missions							
			Maxi	Mini	Micro	Search	Recon-naissance and Mapping	Rubble Removal	Structural Inspection	Victim Recovery	Direct Intervention	Foren-sics	Demon-stration
2007	Crandall Canyon Mine (USA)	Inuktun Mine Cavern Crawler			X	X	X						
2007	Berkman Plaza II (USA)	Inuktun VGTV Xtreme			X	X	X		X			X	
		Active Scope Camera			X	X			X			X	
2009	Cologne Historical Archives (Germany)	Inuktun VGTV Xtreme			X								
		Active Scope Camera			X								
2010	Wangjialing Coal Mine (China)	Unknown		X									
2010	Upper Big Branch Mine (USA)	Mine-permissible ANDROS Wolverine		X									
2010	Prospect Towers (USA)	Inuktun VersaTrax			X	X	X						
		Inuktun VGTV Xtreme			X	X	X						

Note: Shading indicates a robot was on site but not used.

Table 3.1 (continued)

Year	Event	Robot	Maxi	Mini	Micro	Search	Reconnaissance and Mapping	Rubble Removal	Structural Inspection	Victim Recovery	Direct Intervention	Forensics	Demonstration
2010	Pike's River Mine (New Zealand)	Unknown bomb-squad robot	X			X	X						
		Western Australia Water Company custom pipeline inspection robot	X			X	X						
2011	Christchurch earthquake (New Zealand)	iRobot Packbot		X					X				
2011	Tohoku earthquake (Japan)	KOHGA3		X					X				
		Kenaf		X									X
		Quince		X									X
2011–present	Fukushima Daiichi (Japan)	Teleoperated construction equipment	X					X					
		QinetiQ Talon		X			X	X					
		Brokk	X					X					
		iRobot Packbot		X			X				X		
		Quince		X			X						
		JAEA-3	X				X						
		iRobot Warrior	X					X					

Note: Shading indicates a robot was on site but not used.

(NIOSH) have been moving toward making man-portable robots such as the iRobot Packbot and the Sandia National Laboratories Gemini robots mine-permissible. Likewise, the bomb squad robots used for search and later the large pipeline inspection robot used for remediation at the Pike River Mine were ad hoc adaptations of equipment designed for other purposes. Thus, the frequency of use of a maxi robot for mine disasters is less about how well that size robot fits and more about availability. Fukushima Daiichi was hopefully a rare event, and the Institute of Electrical and Electronics Engineers (IEEE) Robotics and Automation Society decided in 2012 to treat nuclear robots, whether for daily operations or for recovery and remediation, as an area distinct from safety, security, and rescue robots.

3.4.1 Performance

The performance of a UGV can be measured in terms of *mission success, time in the area of interest, setup time, errors,* and *human to robot ratio*. Mission success for restricted mobility venues is generally to get sensor coverage of the volume of interest beyond what a camera or range imager on a borescope or wand could do. Mission success for human-habitable or exterior spaces is to accomplish tasks that a human can do, preferably at the equivalent speed.

UGVs have been successful, though they have not made any "live saves"; instead, the value has been in going to places where humans cannot fit or are too risky. UGVs have found victims at the site of the World Trade Center (Murphy, 2003) and were used at only three other events where people were actually missing (Berkman Plaza II, Midas Gold Mine, and Crandall Canyon Mine). At the Berkman Plaza II collapse, the robot was able to search areas that responders could not, but the victim's body was discovered after a few days entombed in concrete through responders' sense of smell. At the Midas event, the robot was looking in the right area but didn't have sufficient illumination to see the body at the distance. At the Crandall Canyon Mine event, the robot had limited mobility in the terrain, and the upper and lower video cameras were rendered nearly useless by mud and dripping drilling foam and groundwater. UGVs have reduced risk to responders and sped up the search process by searching places that were dangerous or impossible to investigate with traditional search tools. At Fukushima Daiichi, robots cleared the way for professionals to reach the reactor units to accelerate mitigation and determined what areas posed acceptable or unacceptable risk to humans. UGVs have provided structural engineers with a new source of information for making decisions and for gathering forensic data.

The *setup time* that is tolerated by the responders and the *time in the area of interest* depend on the situation. For man-packable robots, there is an expectation that the robots be as easy to set up and use as other technical tools, which is on the order of a few minutes. We documented responders at the World Trade Center collapse walking away if the robot was turned on but was not fully functional and displaying video within 1.5 minutes (Casper and Murphy, 2003). For the restricted maneuverability environments that we have worked in, such as the World Trade Center, La Conchita mudslides, the Berkman Plaza II collapse, and so forth, the robots were rarely used for more than 20 minutes at a time because the voids did not extend more than 30 meters. In addition during searches, there is significant time pressure to finish up and move to the next void. At the mine disasters that we have participated in or studied, robots worked for many hours on a single run and sometimes for close to 24 hours at a time. Part of this is the long amount of time it takes for a robot to move from the point of entry to the real area of interest. Another reason is that the UGVs have been used for recovery operations, and there is less time pressure to move quickly. Tethered robots can just stop and wait in place because the tether provides power while the operators take a rest break (which we did at the Midas Gold Mine response) or even get a good night's sleep (which happened at Crandall Canyon Mine).

Errors are discussed in section 3.4.2, but it is worth emphasizing that terminal failures may interfere with normal rescue and recovery operations. If a UGV fails, it may block the path of human responders or of other robots, which happened at the Pike River Mine, or block air or sampling passages, which happened at Crandall Canyon Mine but fortunately at the end of the response.

3.4.2 Mission Failures

Mission failures were encountered during one or more runs at nine events: the World Trade Center disaster, the La Conchita mudslides, the Excel No. 3, DR No. 1, McClane Canyon, Sago, Crandall Canyon, and Pike River Mines, and Fukushima Daiichi. However, only five UGVs have been lost: a wireless Solem at the World Trade Center, the Inuktun Mine Cavern Crawler at Crandall Canyon Mine, the two bomb-squad robots at Pike River Mine, and Quince at Fukushima Daiichi. Using the failure taxonomy from chapter 1, the causes for the mission failures can be further explored. The most common source of terminal mission failures in UGVs has been human error (5 times), followed by mobility problems (5 times, but 3 were with the same robot class and the other 2 involved stairs), environmental factors (3 times), fouling of sensors (2 times), loss of wireless communications (1 time, but this was also the only documented wireless use), power (1 time), and no control failures. Nonterminal failures were much less frequent, with one due to a short circuit and two due to operator fatigue.

3.4.2.1 Terminal Mission Failures Only three of the mission failures were due to *external environmental factors*. At the DR No. 1 Mine, a rock fell and damaged the fiber-optic communication to the V2 robot, ending the mission. On the last run at Crandall Canyon Mine, the borehole collapsed on the robot as it was exiting the borehole, and the robot was lost. However, this was not truly a mission failure as the robot had provided all the data it could. At the Pike River Mine, two UGVs were lost in a methane explosion.

Five mission failures have been the result of *mobility problems*, though this may be because operators are conservative and bring the robot back rather than risk navigating through uncertain terrain. Three of these failures were with either a Inuktun VGTV or VGTV Xtreme, essentially the same class of polymorphic robot with open tracks. The tracks came off a VGTV during one run at the World Trade Center, probably due to extreme heat softening the tracks. The robot was dragged out by its tether and a spare track put on. At the La Conchita mudslides, the tracks came off the Xtreme in 2.17 minutes in mud on its first run and in 7 minutes on shag carpeting when it was searching a house on its second run. In both cases, the robot was still within sight of the operators, and they could use a stick to fish out the track (Murphy and Stover, 2008). The responders terminated the use of the robot because of lack of reliability. This could argue for not using the Inuktun micro-class robots, but there appears to be no other waterproof robot that size with two-way audio. Two failures involved stairs at Fukushima Daiichi, where the original iRobot Packbots were unable to climb the wet, metal stairs, and Quince attempted to climb stairs but had to back down after discovering a landing that was smaller than the construction drawings indicated.

Loss of *wireless communications* has been the source of only one mission failure and has been discussed as a problem at Fukushima Daiichi (Kawatsuma, Fukushima, and Okada, 2012) where the robots used fiber-optic cables if working beyond line of sight. In the known mission failure case, it was the Solem variant of the QinetiQ Talon being used at the World Trade Center response at Building 4 (Murphy, 2003). The robot lost communication on its return from inspecting a void, again good timing. All other disasters have used robots that either had permanent tethers or had optional fiber-optic cables to ensure connectivity. However, the potential for wireless communication appears in the offing as Fukushima Daiichi is reportedly being outfitted with repeaters to allow reliable wireless communication; robots are currently using fiber-optic cables.

Sensing, specifically fouling of cameras, has caused two mission failures. At the Crandall Canyon Mine disaster, the Inuktun Mine Cavern Crawler had to be withdrawn from the borehole before it entered the mine. The video cameras had become coated with drilling foam and mud from the groundwater seeping through the uncased hole. The cameras were wiped off and the robot reinserted, and fortunately the cameras were not again fouled. An iRobot Packbot was unable to enter Fukushima Daiichi Unit 2 because the camera fogged up due to the high humidity (Kawatsuma, Fukushima, and Okada, 2012).

Loss of *power* has resulted in one mission failure, though it is not clear if this was really a mission failure or allowing the robot to die in place. In this case, one of the two New Zealand Defense Force robots at the Pike River Mine ran out of power, but fortunately in an open area where the other robot could maneuver around it.

The biggest source of errors leading to mission failure (5 times) has been ascribed to *human error*. A Talon robot got stuck and was almost abandoned at the World Trade Center. The tether for the V2 mine rescue robot was severed at two different mines when the operator drove over the fiber-optic cable. Likewise, the tether for Quince at the Fukushima Daiichi incident became stuck and eventually snapped, leaving the robot in a place where it could not be recovered (Kawatsuma, Fukushima, and Okada, 2012). At the Sago Mine disaster, the V2 robot was being driven along a narrow-gauge railroad track and wandered off course and fell over when the operator became distracted by performing a general reconnaissance of the mine; fortunately, the robot did not block the path of the human responders (Murphy and Shoureshi, 2008; Murphy et al., 2009a). As noted in chapter 1, it is not clear that these instances of human error reflect the error of the human operator or the error of the designer in expecting the operator to be able to perform significantly complex tasks under unfavorable conditions.

3.4.2.2 Nonterminal Mission Failures   There have been three nonterminal failures where the mission was able to continue; two of these highlight how fatigue leads to errors. In addition, control issues resulted in one temporary mission failure. One of the New Zealand Defense Force bomb-squad robots apparently shut down after being exposed to water leaking from the mine roof; however, the operators were able to restart it the next day and move it out of the way where it then died in place due to lack of power. In two cases, the operators got the robot stuck and had to walk away before successfully resuming operations—illustrating the role of human factors. During the last week of the World Trade Center recovery operations, a Talon robot being used to examine the basement slurry wall became stuck and could not be dislodged. The operators decided to abandon the robot but came back later in the day and quickly freed it. At the Crandall Canyon Mine disaster, the operators could not get the robot to reenter the borehole through the wire mesh. After more than an hour of trying, the decision was made to stop and get some sleep. The next morning, the robot reentered the borehole on the first try.

3.4.3   Tethers and Performance

Tethers have been the source of four mission failures and have been a navigational impediment when used for exploring human-habitable spaces (Micire, 2008; Murphy and Stover, 2008; Murphy, 2010a; Kawatsuma, Fukushima, and Okada, 2012), so they seem a likely candidate for reengineering. However, power tethers reduce the size of a robot allowing it to enter smaller restricted maneuverability regions, and belays are essential for vertical, tortuous environments. Thus, it is helpful to consider the advantages, disadvantages, and strategies for using tethers.

Tethers have at least five advantages. They permit real-time, continuous communications with the robot, regardless of the operational envelope. They can reduce the size and cost while extending the operational time of a robot by moving heavy batteries and computing to the OCU. Tethers can serve as a belay for lowering and raising a robot as it goes through vertical regions and as a safety rope for recovering a stuck robot.

Tethers also present at least three disadvantages, beyond the additional demands on the robot to drag the tether and work around the impaired control dynamics. They require extra effort on the part of the operator to keep up with the tether, even with automatic take-up reels; indeed, our work recommends a person be dedicated to tether management if possible (Burke et al., 2004; Murphy, 2004a; Murphy, Burke, and Stover, 2006; Burke and Murphy, 2007). Even so, tethers tangle and break. In horizontal traverses, the robot has to have sufficient power to drag the tether over the desired distance.

The data from the World Trade Center suggests that for regions with high degrees of verticality and tortuosity, a tether can actively help with navigation more than it hinders navigation. Consider that, on average, the tether manager had to pull, flip, or otherwise manipulate the tether 7.75 times per drop (about 1 time per minute) to keep the tether from getting tangled. But the tether manager manipulated the tether an average of 9.25 times per drop (about 2 times per minute) actively to assist with navigation, such as helping right the robot, pulling it up to allow the operator to try another pass at getting around an obstacle, and so forth.

Here are three recommendations for designing and deploying tethers:

Design the tether to serve as a safety rope. Having a tether and a separate belay or safety rope impedes the mission. The communications tether was cut at the DR No. 1 Mine reopening because the operator was distracted by keeping up with the separate safety rope (Murphy and Shoureshi, 2008). A separate belay had to be attached to the Inuktun VGTV Xtreme at the Midas Gold Mine response, presenting two tethers to coordinate. The communications and power tether had been changed by the manufacturer to a lighter, more flexible cable in the hopes of improving mobility; instead, the cable could no longer support the weight of the robot. Figure 3.9 shows the awkward arrangement at Midas Gold Mine.

For horizontal regions, treat the tether in the way that firefighters treat fire hoses. Firefighters have developed numerous strategies for running fire hoses inside buildings. For example, firefighters are taught to go straight past a turn and then double back, thereby creating slack in the fire hose rather than trying to pull it around the corner of the turn. Chief John Norman, Fire Department of the City of New York, has a classic textbook filled with such strategies (Norman, 2012).

Assign a dedicated person to be the tether manager, and use shared visual communication to help coordinate the activities of the operator and the tether manager. Based on our deployments and our numerous exercises at high-fidelity disaster sites such as Disaster City®, the NASA Ames DART site, and demolished buildings in Florida, Connecticut, and Kansas, overall performance is better—up to nine times better (Burke and Murphy, 2004; Burke et al., 2004; Murphy, Burke, and Stover, 2006; Burke and Murphy, 2007)—if there is a dedicated tether manager. The need for a tether manager doesn't show up when testing robots in benign office environments; this is a case where the relative lack of field work has hurt development of reliable and useful robots.

3.4.4 Human–Robot Ratio

The current best practice is a human–robot ratio of 2:1, where an operator and response expert work together using a coordination protocol, such as **localize, observe general surroundings, look specifically for victims, report (LOVR)** (Murphy, 2004a), and a shared visual display. In all five of

Figure 3.9
Robot with a belay and tether being lowered into void at the Midas Gold Mine response. The tether is hanging down on the right. The apparatus with two round wheels is an exterior light and camera lowered behind the robot to help see its position.

the terminal mission failures due to human error, the operator was working alone and lost track of either the tether or the vehicle. The current best practice for a UGV traversing a highly vertical or tortuous region is a human–robot ratio of 3:1, where a third person is added to serve as the tether manager (note at the Midas Gold Mine response where the robot had a tether and belay, we used two tether managers). A 2:1 ratio is acceptable for search and rescue because responders work using the buddy system; even if the robot could be used by a single person, there will still be a second person present. As we note in Murphy (2011a) and Murphy and Burke (2008), as connectivity increases, more experts out of the hot zone will be able to see the information from the robot and use it for their own roles.

The high incidence of human error leading to terminal and nonterminal failures raises the question of what is the underlying cause of human error and how to address it. Our work starting from our first field exercises with Florida Task Force 3 in 1999 noted that human–robot interaction was a major problem (Casper and Murphy, 2000), and this has been borne out on every deployment that I have witnessed or directly analyzed. Our conclusion from the World Trade Center deployments was that human–robot interaction was *the* major barrier to effective use of man-packable robots for urban search and rescue (Casper and Murphy, 2003; Murphy, 2004a).

The assumption of one operator for one robot is the most significant contributor to human error in UGVs. Our observations and research indicates that the OCU interfaces and concepts of operations are designed and sold for one operator to run one robot, as if one person can effectively drive and look around at the same time. In fairness to the companies, the largest purchaser of the robots, the U.S. military, originally stated that robots had to have a 1:1 ratio in order to be cost-effective without any consideration of mission performance. But a growing body of research, both ours (Burke and Murphy, 2004; Burke et al., 2004; Burke, 2006; Burke and Murphy, 2007) and that of the U.S. Army (Cosenzo et al., 2006), shows that this is not effective in demanding navigational environments such as cluttered or restricted maneuverability regions. Research conducted by my doctoral student, Jenny Burke, showed that two responders working together with a simple coordination protocol were nine times better in terms of performance and overall situation awareness than a single person or two people working ad hoc (Burke and Murphy, 2004; Burke et al., 2004; Burke, 2006; Burke and Murphy, 2007).

From a cognitive engineering perspective, a single operator makes mistakes because of:

Workload overload due to multiple roles The role of being the driver or operator requires not only navigating through complex regions but also being aware of the health and state of a robot with a relatively low mean time between failure (MTBF) rating; this is a full-time job. Meanwhile, the role of the mission specialist or problem holder (i.e., the person acting as a response expert) is also demanding. The mission specialist may be looking for cracks less than 0.25 meter wide, pools of blood, scraps of clothing, and so forth, in partial darkness and from unusual viewing angles. It is hard to do both roles at the same time.

Workload overload due to physiologic and psychological stressors As documented at the World Trade Center (Murphy, 2004a) and confirmed on our subsequent deployments, responders go the first 48–52 hours without sleep, relying only on infrequent naps. The urgency and inescapable emotional impact of a disaster also add stress. Working with robots is even more stressful because they are infrequently used and because they are highly visible to responders and the media, and thus perceived failures can affect the adoption of robots.

Perceptual impoverishment of remote presence   People are working through the robots and through that mediation are losing important perceptual cues. I insist on two-way audio on robots not only for the case of communicating with a survivor but also to hear the sounds that the tracks are making as a clue for navigation. Audio helps answer questions such as: Are the tracks moving at all, did it lose power? Are the tracks spinning in place and can't get traction? Are the tracks spinning but no sound of contact, suggesting the robot is high-centered? The loss of visual cues is subtle but real. A robot camera transforms what would normally be a three-dimensional, multiresolution visual experience into a two-dimensional, single-resolution experience. On the surface, people appear comfortable with two-dimensional imagery, but decades of human factors research suggests otherwise (Tittle, Roesler, and Woods, 2002).

Poor user interfaces in terms of physical design and display content   OCUs are in the difficult situation of needing to be used in bright sunlight, total darkness, and all conditions in-between; screen glare and visibility problems persist. Our work, initiated by Josh Peschel for his doctoral thesis, has shifted to handheld tablets for at least the mission specialist so that the mission specialist can move the device to a more favorable viewing angle. User interface content is often designed for a robot operator or to provide diagnostics used by a designer; these artifacts add clutter and can distract the operator or mission specialist. Many of the tasks are visually directed, such as looking at the surroundings for small items or even finding a thin fiber-optic cable. Yet there is a tendency to divide up a small screen into smaller windows to allow more data to be visible, reducing the actual ability reliably to see what's in those windows or on the screen.

Add another camera syndrome   Designers often add another camera to provide more information to the operator as a workaround to integrating sensors. In the case of Quince, a rear-facing camera was added to view the tether and tether reel, a fixed-forward camera was added to supplement the "main" pan, tilt, zoom camera mounted on a mast to get a broader overview of the environment, a camera was added for the manipulator arm, and a camera was added to read sensors attached to the robot (Nagatani et al., 2013). This meant that the operator could have to contend with five camera windows plus eight other vehicle health and mapping windows on the display.

3.5   Selection Heuristics

In my experience, the *two biggest factors in selecting a robot with a suitable payload for a mission are the scale (does the robot fit?) and traversability (can it maneuver?)*.

These two factors can be put together to form a **mission environment plot**, shown in figure 3.10. The right-hand portion of the plot reflects fairly benign areas where humans could operate. The central area indicates where a person can still move about but is restricted in some way by needing special skills such as technical rope handling, requiring a permit to enter the space, or working in cumbersome personal protection gear. The left-hand regions are too small for humans, and as the spectrum goes to the right, the environment requires even the smallest robots to burrow.

The general rule of thumb is as follows: *Use a tethered man-packable robot for darkest regions, a man-portable robot for mid-range regions, and a man-portable robot for lightish regions, unless environmental constraints or the mission require a heavier robot.*

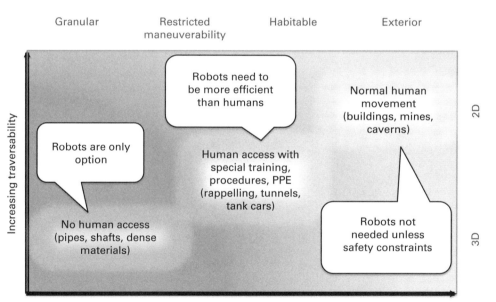

Figure 3.10
Space of mission attributes as a function of scale and traversability.

This rationale is derived from the mission environmental plot. The plot reflects the general operational envelopes for a UGV, as shown in figure 3.11. Panel B of figure 3.11 illustrates that even if a structure is the right scale for human movement and the critical accessibility structure of the stairwell is in place, responders will not enter it until it is shored, thus it falls in a more middle than light area. Because the responders cannot enter due to safety reasons, the region is a candidate for a mobile robot capable of handling a highly cluttered horizontal region and able to climb or descend stairs between regions. (The openness of the remaining structure suggests that a small UAV might be useful; UAVs will be covered in chapter 5.) As the scale and traversability of a region decrease, there is point where a human simply cannot fit, and an alternative such as a robot is a physical necessity. Panel C of figure 3.11 represents one such point, a highly vertical region, where an Inuktun Micro-Tracks is being lowered down a drain pipe at the World Trade Center complex. Panel D of figure 3.11 represents a horizontal regime that is equally hard but for a different combination of scale and traversability: a pancake collapse with 3–5 inches of clearance between the slabs.

The heuristic is supported by the historical data. Returning to the data of section 3.4 on where UGVs have been used, a cluster plot of where the

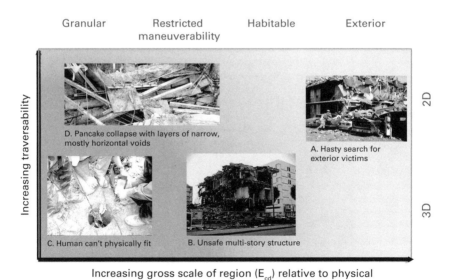

Figure 3.11
Examples of types of operational environments.

different types of mobile robots have actually been used can be superimposed on the plot (figure 3.12). Maxirobots such as the ANDROS F6A and man-portable robots such as the QinetiQ Talon are used extensively for exterior operations and in human-habitable spaces. Maxirobots have been used for mine disasters and for experimental cleanup operations at Fukushima Daiichi. But man-portable robots, which are slightly smaller, much lighter, and also have a manipulator arm, were used for the bulk of Fukushima Daiichi operations and were used to assist maxirobots in clearing debris from roads and covering radioactive material with dirt. The continuing investment in man-portable robots by the U.S. military suggests that this category of robot will become more agile (and autonomous) over time and may replace the maxi class. Man-portable minirobots such as the iRobot Packbot with manipulator arms were used less frequently but have been the primary workhorse for the Fukushima Daiichi nuclear event. Being able to climb stairs and high obstacles is a major capability. The man-packable microrobots such as the Inuktun series have been used most frequently; their sweet spot is regions where human (or larger robot) movement is physically restricted or impossible, especially building collapses.

The relative separation of the clusters is indicative of design decisions that match a robot to an operational environment. However, the clusters

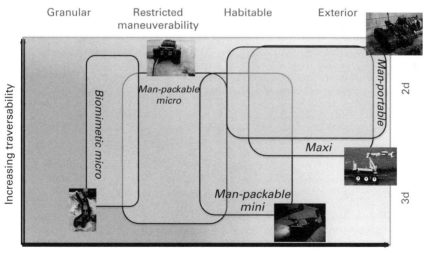

Figure 3.12
Superimposition of what environmental regions have been serviced by what type of currently available robot.

do overlap somewhat, showing that non-navigational constraints and other considerations impact the selection of a robot. For example, the V2 maxirobot has been used for numerous mine disasters because it is the only mine-certified robot in the world; however, as noted earlier, MSHA would like to move to a man-portable or man-packable minirobot.

3.6   Surprises, Gaps, and Open Research Questions

UGVs are arguably the most mature modality for rescue robotics, yet there are still surprises, gaps between what is possible and what is actually implemented, and open research questions. Despite having been the first modality to be used for disasters, rescue is unfortunately different enough from military operations that more work needs to be done.

3.6.1   Surprises
Probably the biggest surprise about UGVs is that *all robots have been tele-operated*. One reason is that all but three robots (KOHGA3, Quince, and JAEA-3) are commercial vehicles used for military operations, public safety, and pipeline inspection applications. There has been no pull from these customers for autonomy. Indeed, as noted in the recent Defense Science Board study on unmanned systems (Murphy and Shields, 2012), the U.S. Department of Defense insisted that even basic autonomous capabilities such as self-righting, a capability developed under DARPA funding, be stripped from platforms.

Another surprise is that *all but five of the 19 models of UGVs actually used were commercial off-the-shelf (COTS) technology*. The exceptions were Souryu (Niigata Chuetsu earthquake), Inuktun Mine Cavern Crawler (Crandall Canyon Mine), KOGHA3 (Tohoku earthquake), and the JAEA-3 and Quince (Fukushima Daiichi). It should be noted that the Inuktun Mine Cavern Crawler was in some sense COTS technology; it was configured from parts designed for original equipment manufacturers to use in building water-proof robots. The use of a COTS robot is a reality of economics; rescue robotics is not a sufficiently large market for the creation of robots specifically for disasters. One manufacturer refused to build a custom robot to CRASAR specifications at any price, saying that the task would completely absorb the company's best designer and take him away from developing robots for applications with a larger potential repeat sales volume.

Only one wireless robot has been used in the initial rescue of a disaster, and it was lost and never recovered due to lost connectivity. The only robot that depended on a wireless network was the Foster-Miller Solem used at the

World Trade Center. It lost communications in a void under Building 4 and was never recovered.

The best practice is a human–robot ratio of 2:1 in general and 3:1 for disasters with extremely restricted maneuverability requiring tether management. These best practices are for teleoperated robots and may change as more autonomous capabilities are added. In the meantime, focused studies in human–robot interaction for urban search and rescue convincingly show that 2:1 is far superior to a 1:1 ratio. Human error from a single operator (1:1) has been the largest source of mission terminal and nonterminal failures, emphasizing the need to do things differently.

Underground mine disaster responses are the most frequent users of rescue robots. Twelve of the 23 deployments or prestaging of robots were at underground mines.

Tethers are problematic for habitable regions but essential for restricted maneuverability regions. Robots at 14 of the disasters were known to be tethered, either to ensure communications or because they required a power/comms tether; the actual number may be higher.

Robots of any size should be at least waterproof, have two-way audio, sharable displays, and a secure attachment point for a safety rope. Robots need to be waterproof because there will likely be water at some point in the mission, either groundwater, sprinkler systems, or rain, and because they have to be decontaminated. Two-way audio is needed for search and surrogate missions, but more importantly audio is needed to hear terrain cues. Sharable displays enable the best practice of a 2:1 or 3:1 ratio. Given that even maxirobots have required a safety rope at some point, all robots need a way to attach one.

Power has not been a major mission-failure mode. Battery power needs to be matched to the robot and application. For example, an Inuktun VGTV class robot at structural collapses searched voids in less than 20 minutes per void, but a single run for the same class of robot at a mine rescue took close to 12 hours.

Disasters have not used coordinated multirobot teams. While multiple robots have been present at seven deployments, they are typically used independently and not formally coordinated. Robots may come from different agencies and are configured to work as if they are the only robot present. This can lead to problems in even turning on the robots if they use wireless, as the network can become saturated; we have witnessed and reported these types of coordination problems in Kawatsuma, Fukushima, and Okada (2012). But even if multiple robots are expected to function as a team, there is no special interface. For example, two sets of multirobot teams were used at Fukushima Daiichi. A Packbot was used to help another Packbot for opening doors, and a Talon robot was used to help a Bobcat move debris, but these were teleoperated and there were no special interfaces or devices to help the operators to interact with each other.

3.6.2 Gaps

The gaps in UGVs remain similar to those we identified in 2000 (Murphy et al., 2000), 2005 (Carlson and Murphy, 2005), and 2006 (Murphy and Stover, 2006a) and later confirmed by other experiences in Micire (2008). The biggest gaps are better user interfaces, increased reliability in extreme conditions, support for "plug and play" sensors, and interoperability, including access to APIs; none of these require research but do require industrial development. Progress has been made in transportation, the ability to decontaminate the robots, and compliance with ISO 9000 standards on quality control. Robots such as the Packbot now come with CBRNE sensor packages that can be added as needed. Our work with hazmat responders has produced a list of requirements specifically for that application (Murphy et al., 2012b).

Based on our 2006 gaps analysis (Murphy and Stover, 2006a) and the experiences since then, the recommended functionality of a rescue robot is to:

• *Carry and record data from basic sensors*: color video with zoom, infrared video, two-way audio, low-temperature lighting (halogen lights can ignite explosive atmospheres), environmental temperature, and allow all data to be recorded and to be viewed on external, secondary OCUs.
• *Allow the camera pan and tilt to be independent of the platform*, as often manufacturers simplify the design and increase reliability and waterproofing by fixing the camera on the platform, expecting the platform to pan for the camera. In restricted maneuverability regions, the platform cannot turn more than a few degrees in either direction, thus constraining the camera to face forward.
• *Support essential functional attributes*: especially, be waterproof for decontamination, be invertible or self-righting with the payload intact and functioning (some manufacturers demonstrate running upside down or self-righting with a bare-bones platform, not with the sensor package essential to the mission), and allow the attachment of a safety rope, and if a power/communications tether is used, allow a hot disconnect/reconnect.
• *Minimize dust/noise generated by locomotion* so as not to interfere with cameras, rangers, or audio sensors.

• *Provide self-cleaning sensors or sensor housings* so that sensors are not fouled by mud or water.
• *Provide a smooth "hangless" package* to prevent a robot from getting stuck or snagging exposed wiring; robots with sensor masts are particularly vulnerable to having the masts knocked off.
• *Be a visible yellow or orange color* to make it easier to spot in the rubble and to verify decontamination.
• *Move to the last point(s) of connectivity* if wireless is lost.

3.6.3   Open Research Questions

Open research issues remain in autonomy for navigation and manipulation, human–robot interaction, and new mechanisms able to penetrate extremely restricted mobility or granular regions.

Reliable autonomous capabilities are needed, especially obstacle avoidance and, as noted in Micire (2002), maintaining the center of gravity as the robot negotiates obstacles. It is very difficult for even experienced operators to estimate depth and sizes of obstacles and relate those to the clearances and capabilities of the platform.

Navigational route planning is not critical. The standard operating procedure is for robots to use the same right-wall–following strategy as human responders to ensure repeatability in habitable environments. In restricted maneuverability regions, the robot usually has no choice of turns, but if there are openings, the right-wall–following approach is suitable.

Simultaneous localization and mapping (SLAM) would be valuable for rubble or voids where any a priori knowledge has been obviated by the disaster, such as a building collapse; otherwise, localization and mapping by the human operator is fairly straightforward. A particularly daunting challenge for SLAM is how to localize and map in highly irregular vertical regions where the robot may be rappelling down and not guaranteed to be in contact with the terrain at all times.

Manipulation, either for navigation, investigation, or intervention, is becoming increasingly important. Current teleoperated manipulation requires two robots, one to provide an external view of the manipulator relative to the target object. It is at best painstaking and slow. Autonomous capabilities for handling at least basic accessibility elements without the need for a second robot would provide performance improvements and reduce the strain on the operator.

Human–robot interaction, especially conceptualizing the human–robot team as a joint cognitive system where capabilities may be delegated or shared, is a key area of research. The high number of mission failures attributed to human error reinforces the need for better human–robot interaction.

While the robots are fairly competent for most operational environments, extremely restricted and granular regions are too difficult for COTS robots. Advances in snake and burrowing robots will hopefully be able to move forward literally in these regions.

3.6.4   Rating of Barriers to Implementation
It is interesting to note that a November 2012 poll of the members of the IEEE Robotics and Automation Society's Technical Committee on Safety, Security, and Rescue Robotics rated the barriers in implementation of robots for the public safety sector as:

1. Human–robot interaction
2. Perception
3. Control and planning (tied)
4. Test and evaluation (tied)
5. Mechanisms
6. Modeling and simulation

3.7   Summary

UGVs provide remote presence and intervention capabilities on the ground, either in interiors or exteriors. The types of tactical UGVs used for rescue have historically been called *mobile robots* and can be partitioned by size and portability into three major categories: maxi, man-portable, and man-packable. Maxirobots work in exteriors and spaces that a human could maneuver through, but are heavy and cumbersome. Man-packable robots are small, easy to carry to the point of ingress, and can work in areas of interest too small for humans and dogs. Man-portable robots occupy the middle of the size and portability spectrum: they work in the same scale as a small human but can be carried to the point of ingress.

A UGV system has to function in three distinct environments. One is the robot's operational envelope, which is the region(s) that the robot works in and which may present navigational challenges due to the relative scale of the region, its traversability, and non-navigational constraints such as explosive atmospheres. The second is the point of ingress or egress, which historically has been the same point: The robot has to get to and enter the region of interest somehow. The third is the operator's environment, or where humans will be working. If the robot system does not fit any of these three environments, the robot will not be used. Comparative case studies of how robots were not used at Hurricane Charley but were used at Hurricane Katrina illustrate the influence of the environments on the choice of robots.

UGVs have been used primarily for reconnaissance and mapping missions, often combined with search or victim recovery for interiors. The most common events have been mine disasters, followed closely by structural collapses from meteorological events, industrial accidents, or terrorism; however, the Fukushima Daiichi nuclear emergency engaged many types of mobile robots for both exterior and interior operations. UGVs have not made a "live save" but have increased the efficiency and coverage of rescue and recovery operations while reducing the risk to responders.

These deployments lead to recommendations on robot platforms, capabilities, human–robot ratio, and future research. It is recommended that a robot be waterproof, a visible yellow or orange color, be either invertible or self-righting, and allow the attachment of a safety rope. It should carry a basic payload of a color video camera with the best optics and zoom possible, two-way audio, low-temperature lighting (halogen lights can ignite explosive atmospheres), and environmental conditions. The camera should pan and tilt independently of the robot platform. The entire platform and sensors should form a smooth package to prevent snags or hanging on irregular obstacles. The robot should not stir up dust or make loud noises, and sensors should be self-cleaning as mud and dirt often coats the robot. If the robot is wireless, it needs a return to home or to the last point of connectivity functionality. To overcome the prevalence of human error as a major cause of mission failures, a 2:1 human to robot ratio is recommended for mobile robots in general and a 3:1 ratio for operations in regions of highly restricted maneuverability where the tether is actively affecting, either assisting or impeding, navigation. Tethers should be strong enough to serve as a safety rope, and procedures should mimic how firefighters run fire hoses in complex interiors. Open research questions are autonomy for navigation and manipulation, better human–robot interaction, and new mechanisms able to penetrate extremely restricted mobility or granular regions. The need for manipulation will continue to grow as many missions and tasks involve navigational, investigatory, or intervention manipulation: *search, rubble removal, in situ medical assessment and intervention, medically sensitive extrication and evacuation of casualties, adaptive shoring, victim recovery,* and *manipulation of key components.*

The chapter presents a novel relational plot of the two main attributes affecting the selection of a UGV for a particular operational environment:

the gross, or overall, scale of the region (exterior, habitable, restricted maneuverability, and granular) and the traversability of the region (tortuosity, verticality, surface properties, severity of obstacles, and number of accessibility elements). The plot helps to explain why certain sizes of robots are used for specific events and to project what type of robot will be needed for a new event. The plot also highlights the lack of capability for operating in granular regions, where a robot would essentially need to burrow.

The most common modes of failure are human error, followed by mobility problems, environmental factors, fouling of sensors, and loss of wireless communications and power. The reliance of teleoperation for real-time remote presence with no guarded autonomy appears to put too great a cognitive load on the operator, while taskable agency is not desired by the responders. A human–robot ratio of two people to one robot results in a ninefold increase in situation awareness. Tethers are often essential for navigating in complex, vertical spaces and for ensuring real-time communications.

The data on what UGVs were used for what disasters should be thought of as both a temporary guide to selecting currently available technology and a basis for accelerating the development of new robots to fill the gaps. The relational plots show that few robots exist that can handle highly confined, tortuous, three-dimensional spaces, and none can handle granular regions. This highlights the need for more biomimetic platforms. However, the data suggest that better mobility without better human–robot interaction, sensing, and control will not lead to useful systems.

4 Unmanned Aerial Vehicles

Chapter 2 described how unmanned aerial vehicles are a growing segment of rescue robotics. Of the 34 documented incidents where robots have been used, unmanned aerial vehicle (UAVs) have been present at 11, putting UAVs in second place for frequency of deployment. Large UAVs such as Predators and Global Hawks are routinely used for disaster response, but usually tactical responders on the ground do not direct data collection from those strategic assets or have real-time access to the imagery being produced. This chapter focuses on small and micro UAVs that responders can carry with them and the increasing role of these UAVs in disaster response and recovery.

By the end of this chapter, the reader should be able to:

- Describe the two sizes (*small* and *micro*), types (*fixed wing* and *rotary wing*), and styles of control (*teleoperated, delegated,* and *guarded motion*) of UAVs and the pilot primary viewpoint (*heads-up* or *heads-down*).
- Describe the three operational environments (*wide area, local area, interior*) involved in selecting a UAV.
- Describe the common missions and tasks for UAVs.
- Demonstrate familiarity with where UAVs have been used.
- Demonstrate familiarity with the performance of UMVs and their common modes of failures.
- Use the recommended concept of operations by assigning the crew roles and following the operational procedures.
- Match a UAV with the environment and mission using the selection heuristics.
- Show familiarity with the gaps in the technology and the open research questions.

The chapter begins by describing UAVs in general and then focusing on the classes of UAVs that appear to be best suited for organic, tactical use by emergency professionals. UAVs are either like a plane, called **fixed-wing craft**, or like a helicopter, called a **rotorcraft** or **rotary-wing craft**. The chapter discusses the three different operational environments for a UAV. Fixed-wing UAVs are well suited for wide-area operational environments where reconnoitering large geographical areas is of importance. Rotary-wing UAVs are better suited for working near structures and other local-area operational environments and in interior environments. The chapter describes the missions and tasks for UAVs (which currently are limited to being the remote eyes of a responder), where UAVs have been used, their generally successful performance, and their relatively few mission failures. However, deployments and studies indicate an unacceptably high cognitive load on human team members. Our research has created a concept of operations that specifies three crew roles, a set of operational procedures, and a new type of interface to mitigate the cognitive overload. The chapter then returns to a discussion of environments and how they lead to selection heuristics. The chapter concludes with a list of surprises, the implementation gaps from both the responders' perspective and the research and development perspective, and open research questions.

4.1   Types of UAVs

UAVs are often thought of in terms of platform design, though an equally important distinction is the style of control. The physical components of a UAV system consist of the platform, the operator control unit (OCU), and a communications link (often two links, one dedicated to control and the other to imagery). Regardless of platform design or style of control, the U.S. Federal Aviation Administration (FAA) refers to UAVs as **unmanned aerial systems**, or **UASs**, to emphasize that regulations require a human pilot to be engaged with the UAV. The U.S. Air Force refers to UAVs as **remotely piloted vehicles**, or **RPVs**, to acknowledge the role of the human in the ultimate control of military aircraft. For the sake of symmetry with the terms UGV and UMV, which vehicles also have humans in the loop, this book will use the term UAV.

4.1.1   Platform Design

A UAV design falls into one of two categories based on flight mode: fixed-wing craft (i.e., a plane) or rotorcraft (i.e., a helicopter). Fixed-wing UAVs are often hand-launched from a parking lot or a meadow and land by stalling and floating down. They may have a flight endurance of one or more hours. Fixed-wings will often have GPS waypoint navigation and return to home if the signal is lost.

Rotary-wing UAVs may look like either a miniature helicopter or a pizza-dish-like array of rotors. Quadrotors are a popular configuration as pairs of rotors can counteract each other, but the number of rotors varies from three to eight. Rotary-wing UAVs have a much shorter endurance, of the order one-half hour of hovering, which takes a great deal of energy. In general, multirotor UAVs are **stick neutral**; they maintain altitude without input from the pilot and location (if GPS is available). More multirotor models are adding GPS waypoint navigation, return home on signal loss, and auto-land capabilities. Miniature helicopter models tend to cater to the stunt hobbyist and are not necessarily stick neutral.

An *organic* UAV, one that can be carried and deployed directly by a responder, is usually further classified as either being *small* or *micro* in size, though there is some overlap as there are competing classifications within the U.S. military and the FAA (Peschel and Murphy, 2013). For the purposes of this book, a **small UAV** will follow the FAA definition of being under 55 pounds (25 kilograms) and capable of limited altitudes (Sizemore, 2010). A **micro UAV** typically is less than 1 meter in any characteristic dimension and weighs less than 1 kilogram, though in the original DARPA Micro Aerial Vehicle program, which accelerated the development of small UAVs, the characteristic dimension was put at 15 centimeters. In general, a small UAV (abbreviated **sUAV**) can fly at higher altitudes and in stronger winds than a micro UAV (micro aerial vehicle; **MAV**). Small fixed-wing UAVs often have collapsible wings and fuselage to allow them to be easily transported or fit into backpacks. The fixed-wing UAVs typically have a set of fixed cameras; instead of panning and tilting, the operator selects a different camera or flies differently. Rotary-wing UAVs generally allow the onboard camera to tilt as the platform does not tip, but the entire platform usually turns to act as a pan movement. Examples of small and micro UAVs are shown in figure 4.1.

4.1.2 Style of Control
UAVs differ in the style of control, which can be divided into **autonomous capabilities** and **pilot primary viewpoint**. Autonomous capabilities are associated with a platform, while pilot primary viewpoint refers to where the pilot is looking during operations.

Different manufacturers provide evolving autonomous capabilities for navigation, beyond basic flight stability; these fall into three categories, which are not mutually exclusive. One category is **teleoperation**; UAVs may be stick neutral but the human pilot is responsible for all movements and actions. The second category is **delegation**. Other UAVs allow the

Figure 4.1
Examples of micro UAVs: (left) a fixed-wing AeroVironment Raven, (middle) an iS-ENSYS IP-3, and (right) an AirRobot 100 quadrotor.

pilot to delegate navigation functions or the vehicle to assume functional responsibility under well-defined circumstances. For example, many UAV have *automated return-to-home* capabilities upon sustained loss of signal and *auto-land*, either for pilot ease or for low battery conditions. *Waypoint navigation* where a UAV is tasked to visit one or more GPS locations is also common, though these programs do not include obstacle avoidance (also called **sense and avoid**) capabilities, so the pilot is responsible for making sure the UAV is operating in open space. Other capabilities include *return to a previous location and heading*, which helps responders periodically check on a particular area or resume operations at a specific spot. Less common is *autonomous takeoff*, as the vehicle may not have yet acquired a strong or accurate GPS signal, and thus a manufacturer may want a human pilot to maintain oversight. The third category is **guarded motion**, where the pilot directs the vehicle but the vehicle provides some protection. Guarded motion capabilities in UAVs include *staying within a bounding box* and *flying at a specific altitude*, though guarded motion in unmanned systems and spacecraft includes a richer set of aids, including collision avoidance (Pratt and Murphy, 2012).

UAVs are designed to be either primarily **head-up**, where the pilot looks at the UAV during operations with occasional glances down at the display, or **head-down**, where the pilot looks down at the display and rarely at the UAV. The ramifications of using either a heads-up or heads-down style of control on pilot performance and error have not been studied. The heads-up style is an extension of radio-controlled hobby aircraft, where the pilot watches the UAV to direct it through complex maneuvers. The advantage of this **exocentric view** is that the pilot always knows where the vehicle is relative to other objects in the area such as trees and buildings. The disadvantage is that it can be hard to determine the best data collection positions, especially as the UAV works at longer distances and becomes less distinct. The heads-down style relies on an **egocentric view** through the UAV's cameras. This makes data collection easier, but the pilot has to maintain mentally an internal awareness of where the UAV is in relation to other objects in the area. The pilot may have to swap to a heads-up viewpoint for landing and takeoff. Heads-up displays that combine both modes are intuitively appealing, but visual acuity, focus of attention, and ergonomics remain barriers to adoption.

4.2   Environments

A UAV may operate in three environments: **wide-area** expanses, **local areas** near structures or terrain features, and **interiors** of buildings. Figure 4.2 shows this general partitioning. The airspace for wide and local areas may be subject to air traffic control regulations, depending on the country. For example, in the United States as of April 2013, all UAVs unless privately owned must have a certificate of authorization to fly at any altitude, despite hobbyists being allowed to freely fly in most areas under 120 meters.

4.2.1   Wide-Area Operational Environments

Wide-area operational environments, such as surveying a spill or looking for a person lost in the wilderness, are generally covered by fixed-wing

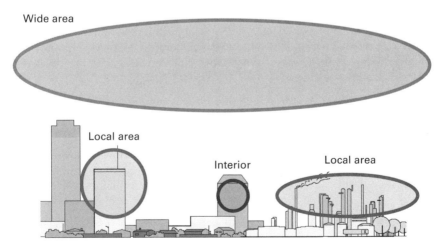

Figure 4.2
A graphical depiction of the three operational environments for UAVs.

UAVs. These miniature planes can circle large areas for hours, easily ranging several kilometers (depending on applicable flight regulations). The operational altitude is between 45 and 120 meters, high enough to avoid collisions with trees but low enough to stay out of normal manned aircraft space. The altitude is favorable for reliable GPS readings and maintaining wireless connectivity with the OCU, though there is little time for recovery from a mechanical failure.

Wide-area operational environments occur in association with meteorological and geological events (such as earthquakes, hurricanes, and flooding), wildfires, wilderness search and rescue, and chemical, biological, radiological, nuclear, or explosive (CBRNE) events. Larger UAVs such as the Predator tend to be used strategically for meteorological events rather than organic UAVs. Small and micro UAVs have been used since Hurricane Katrina, first by us (Murphy, 2006; Murphy et al., 2006c) and later by the U.S. Army. More exciting in terms of UAVs for meteorological events was the use of multiple fixed-wing UAVs during the 2011 Thailand floods to monitor large areas and allow disaster scientists to predict and prevent flooding (Srivaree-Ratana, 2012). The U.S. Forestry Service has been exploring different sizes of UAVs for wildland firefighting, and the Ikhana Global Hawk was used in a major California wildfire (Ambrosia et al., 2010). Small and micro UAVs do not appear to have high utility, as a focus group sponsored by the Association for Unmanned Vehicle Systems International (AUVSI) found that firefighters favored larger systems that could fly large distances at night with specialized electro-optical/infrared (EO/IR) platforms (Murphy, 2010b). The use of small and micro fixed-wing UAVs for wilderness search and rescue has been explored (Goodrich et al., 2008), and some agencies are now using them. UAVs for CBRNE events have been heavily researched in terms of accurate plume tracking, though our work with responders indicates that general situation awareness of what is in the area that might be affected by the plume is of more value (Murphy et al., 2012b). A Honeywell T-Hawk was used extensively at the Fukushima Dai-ichi nuclear incident to conduct radiation surveys and general inspection.

4.2.2 Local-Area Operational Environments

Local-area operational environments are generally occupied by rotary-wing UAVs that can fly at lower altitudes and hover near structures. Our work performing structural inspection in the aftermath of Hurricane Katrina (Pratt et al., 2009) and later with developing a concept of operations for CBRNE events (Murphy et al., 2012b) suggests that the range of altitudes is from 3 meters to about 45 meters. A rotary-wing UAV can go higher, but the specified altitude range seems to give good coverage of a local area of interest. While rotary-wing UAVs can travel several kilometers, the real value appears to be in hovering for examining a situation. Our experiences suggest that a UAV will need to get 2–5 meters from a structure (Murphy, Pratt, and Burke, 2008). This proximity to buildings, train wrecks, or metal structures introduces GPS and control signal multipaths and exposes the UAV to wind shear and difficult-to-model conditions. There also may be visual challenges such as working in clouds of steam or chemicals or at night.

Local-area operations tend to place the OCU and team as close to the area of interest as possible, posing real risks. The team at the Cypriot naval base had to work near a large amount of fuel in the ruptured fuel tanks (Angermann, Frassl, and Lichtenstern, 2012). In both the Cypriot and Katrina responses, there was limited space for the team, and a safety officer had to make sure the pilot did not trip over debris (Pratt et al., 2009; Angermann, Frassl, and Lichtenstern, 2012). Indeed the concept of operations we developed during our work at Hurricane Katrina specified this as a task for the safety officer, and this was the protocol used in Cyprus.

Local-area operations are associated with structural collapses, regardless of cause, and identification of the source and impact of a CBRNE event, such as an oil spill or chemical train derailment. UAVs are beginning to gain traction for prevention of collapses and spills, moving UAVs from a reactive response and recovery resource to a preparedness and prevention role.

4.2.3   Interior Environments

A third operational environment, *interiors* of structures, is rapidly emerging as a key application domain. Micro UAVs offer the advantages of enhanced maneuverability in cluttered spaces, and to date three deployments have involved interiors. Structural engineers used a Parrot AR drone to inspect the Christchurch Cathedral after the Christchurch earthquake in 2011. A team from the University of Pennsylvania demonstrated a Pelican quadrotor UAV mapping the interior of a damaged building on the University of Tohoku campus (Michael et al., 2012). The Natural Human–Robot Cooperation in Dynamic Environments Project (NIFTi) consortium of European universities fielded two custom quadrotors to inspect damage to historic cathedrals in Mirandola, Italy, after the 2012 earthquake (Kruijff et al., 2012).

Flying indoors raises the question of *how do the UAVs get inside rubble?* This is analogous to the point of ingress/egress as a distinct environment for UGVs, as discussed in chapter 3. One answer is that they are carried in by a UGV. In the case of the Tohoku trials, the UAV was carried by a UGV. In Christchurch, the UAV was launched at the opening. Another approach is that the UAV flies from the exterior into an interior space. During our post–Hurricane Katrina inspection work in 2005, our UAV lead, Chandler Griffin, tried to fly into one of the damaged casinos to get a better view. Even with a straight line of sight, the connectivity began to degrade in the twisted metal structure. The UAV was only able to penetrate 20 meters inside, not deep enough to get the needed view. However in 2012, the NIFTi team successfully deployed their custom quadrotor from the outside of a San Francisco church through a door in a gallery and then in the main church and adjoining gallery, completely out of the pilot's view (Kruijff et al., 2012).

4.3   Missions and Tasks

UAVs have been used for only two of the rescue robot missions identified in chapter 1. The two missions have occurred with near equal frequency: *structural inspection*, which has occurred at 7 of the 11 deployments, and *reconnaissance and mapping*, which has occurred at 6 events. The UAVs have been used for visual assessment, either through video or RGB-D cameras, so the only task has been to capture imagery. However, UAVs for actually interacting with the environment are being examined (Peschel, 2012b).

4.3.1   Where UAVs Have Been Used

Table 4.1 summarizes the 11 known UAV deployments previously described in chapter 1. Each row lists the robots used (if known) for each event by size, type, and mission. The data suggest two interesting trends:

Local-area operations appear to be the "killer app."   Only two events used a UAV for wide-area operations. We used a Raven at Hurricane Katrina to determine the situation at Pearlington, Mississippi, and along the Pearl River; these areas were unreachable by truck due to trees in the road (Murphy et al., 2006c). An Elbit Skylark was used at the Haiti earthquake to determine if a distant orphanage was intact. Information about a wide area is needed at the beginning of a response, so if the responders do not have immediate access to a UAV (recall from chapter 1 that it takes an average of 6.5 days before robots are used), eventually they will get the information from manned helicopters, National Guard Predators, or other assets. Wide-area operations may become important when responders have their own UAVs.

Rotary-wing UAVs appear to be the most desired platform.   The preference for rotary-wing UAVs is most likely related to the trend toward local-area applications, which are suited for hovering and flying near structures. However, the general feedback I have gotten is that low-altitude hovering is something that cannot be done with manned helicopters and thus is extremely valuable as a new tool. A manned helicopter or plane or even a larger UAV can provide data similar to a fixed-wing UAV and often with better optics. We saw this at the 2007 Berkman Plaza II collapse where we used an iSEN-SYS IP-3 for forensic documentation of the collapse. A manned helicopter could not get acceptable pictures because the helicopter rotors kicked up dust obscuring key features. Furthermore, the collapse was in downtown Jacksonville, Florida, next to multistory buildings and the St. John's River, creating unpredictable wind gusts and generally unsafe conditions.

4.3.2   Performance

The performance for a UAV is rated by *mission success* and related factors that influence user trust and adoption: *setup time* (Can it be used quickly?), *failures or errors* (Will there be crashes or repeated missions?), and the *human to robot ratio* (How many people have to be on-site? How can cognitive errors be reduced?). Mission success is subjective but usually resulted in documentation of a structural condition (such as the Cypriot naval base, the Christchurch Cathedral, etc.) or social condition (such as the status of Pearlington after Hurricane Katrina or the Haitian orphanage after the earthquake). Data were reported for setup times beyond the approximate 15 minutes we spent at the post–Hurricane Katrina recovery flights (Pratt et al., 2009), though the team at the Cypriot naval base was able to complete 10 flights in 1 day (Angermann, Frassl, and Lichtenstern, 2012) suggesting a similarly short setup time. Failure data were also not directly reported, but what is known is discussed later in this section. The human–robot ratio, when reported, was 3:1 with our flights at the Berkman Plaza II collapse (Pratt et al., 2008), the Westinghouse–Honeywell team at Fukushima Dai-ichi, the Cypriot naval base deployment (Angermann, Frassl, and Lichtenstern, 2012), and the Finale Emilia earthquake deployment (Kruijff et al., 2012) explicitly following the concept of operations we established after Hurricane Katrina (Pratt et al., 2009; Murphy, Pratt, and Burke, 2008).

4.3.3   Mission Failures

Mission failures were reported at four events: Hurricane Katrina, Berkman Plaza II, Fukushima Daiichi, and the Finale Emilia earthquake. Of these, three were terminal and one was nonterminal. Only one UAV has been

Table 4.1
Table of known deployments of UAVs

Year	Event	Robot	Size		Type		Mission		
			Small	Micro	Fixed-Wing	Rotary	Reconnaissance and Mapping	Structural Inspection	Demonstration
2005	Hurricane Katrina (USA)	Like90 T-Rex	X			X	X		
		AeroVironment Raven		X	X		X		
		iSENSYS IP-3	X			X		X	
2005	Hurricane Wilma (USA)	Like90 T-Rex		X		X		X	X
2007	Berkman Plaza II (USA)	iSENSYS IP-3		X		X		X	
2009	L'Aquila earthquake (Italy)*	Custom		X		X			X
2010	Haiti earthquake (Haiti)	Elbit Skylark	X		X		X		
2011	Christchurch earthquake (New Zealand)	Parrot AR drone		X				X	
2011	Tohoku earthquake (Japan)	Pelican		X			X		
2011	Fukushima Daiichi nuclear emergency (Japan)	Custom	X		X		X		
		Honeywell T-Hawk	X			X	X	X	
2011	Evangelos Florakis naval base explosion (Cyprus)	AscTec Falcon		X		X		X	
		AscTec Hummingbird		X		X		X	
2011	Great Thailand Flood (Thailand)	Unknown	X		X		X		
2012	Finale Emilia earthquake (Italy)	NIFTi 1		X			X	X	
		NIFTi 2		X		X	X	X	

reported lost, a Honeywell T-Hawk that made an emergency landing on a reactor building and has not been recovered. The mission failures fell into two categories using the taxonomy from chapter 1: *human error* and *power* (engine) failure.

Human error was attributed to three nonterminal failures caused by the *inability to fuse multiple models of the environment from different frames of reference, general fatigue,* and *lack of situation awareness.* These are described below.

• *Frames of reference* At Hurricane Katrina, while landing in a demolished neighborhood, we crashed the Raven into a set of power lines (Pratt, 2007). The power lines were clearly visible to the pilot and safety officer, and they had remarked upon the power lines as they began to plan the landing. But the power lines were hard to see in the OCU screen, visible in only 15 frames (see figure 4.3 to note the perceptual difficulty). The pilot was not able to maintain overall situation awareness by merging what he had previously seen in his egocentric frame of reference and commented on minutes before with what he was (or was not) seeing from the robot's egocentric viewpoint.

• *Cognitive fatigue* The pilots at the Finale Emilia earthquake had increasingly sloppy landings and broke three rotor blades in collisions with the entryway as the day wore on, presumably due to cognitive fatigue. The report emphasized that even though the flights were of the order 5 minutes long, the pilot was under great stress. The intense stress of flying at disasters has long been noted in our work and was mentioned by the engineer who flew the AR Parrot drone at Christchurch, New Zealand.

• *Situation awareness* The deployment in Mirandola, Italy, also gave evidence of a failure to maintain situation awareness of the UAV, possibly from fatigue or from the more general problem of displaying the robot state to an operator. In that instance, the pilot directed the UAV platform to a spot to inspect a cupola but failed to take into account the position of the camera. Maintaining awareness of the camera relative to the robot is very-low-level situation awareness needed as a building block for the broader mission-oriented situation awareness following Endsley's taxonomy (Endsley, 1988); the failure to maintain this low level in turn suggests that even larger problems with situation awareness were lurking. Four minutes of the 6-minute flight were lost as the pilot had to diagnose and correct why the camera was not providing the expected view. This suggests better displays of robot state or coordination between the pilot and payload operator.

• *Power failures*, specifically engine failures, produced two terminal failures. The iSENSYS IP-3 suddenly lost engine power and was slightly damaged when the pilot executed an emergency landing in the rubble (Pratt et al., 2008). It was recovered, as it was safe to walk on the rubble pile; the UAV was there to provide forensic documentation. As noted earlier, the Honeywell T-Hawk made an autonomous emergency landing on Reactor Building 2 at Fukushima Daiichi. The cause of the engine shutdown could not be determined from the transmission log.

4.4   Human–Robot Interaction

Human–robot interaction in UAVs is especially important as operators cannot stop and think the way that UGV and UMV operators can when a problem arises. The consequences of a UAV failure can be high. The UAV can be lost. It could make the situation worse; for example at Fukushima Daiichi, there was a real concern over the possibility that if the T-Hawk crash landed

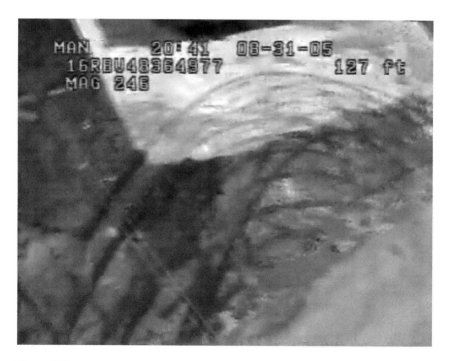

Figure 4.3
A still image from an OCU showing three power lines (lower left) just before the UAV crashed into them at the Hurricane Katrina response.

in the unspent fuel pond, the uranium rods could topple into each other and become an open-air nuclear reactor. The UAV can cause an accident; Angermann, Frassl, and Lichtenstern (2012) note that three members of a UAV team were killed by their UAV during a landing failure.

Using cognitive work analysis and applying principles from human factors, especially lessons learned from the introduction of autopilots into manned aviation, we have developed a concept of operations for small and micro UAVs (Murphy, Pratt, and Burke, 2008; Pratt et al., 2009). This has been used in six deployments: Hurricane Katrina (Pratt et al., 2009), Hurricane Wilma (Murphy et al., 2008a), the Berkman Plaza II collapse (Pratt et al., 2008), at Fukushima Daiichi by the Westinghouse–Honeywell team, the Cypriot naval base (Angermann, Frassl, and Lichtenstern, 2012), and at the Finale Emilia earthquake (Kruijff et al., 2012). The concept of operations consists of three crew roles and a set of operational procedures.

4.4.1 Three Crew Roles

At a minimum, there are three distinct roles and a dedicated person for each role. The *pilot*, also called the operator, is responsible for preparing, flying safely, and maintaining the UAV. The *mission specialist*, also called the payload specialist, stakeholder, or problem holder, is responsible for collecting data. The mission specialist role is usually an active role, operating the sensors or advising the pilot where to fly. The *flight director*, also called the safety officer, is responsible for the overall safety of the team. Consistent with general safety practices, the flight director role must be handled by a dedicated person who has over-watch of the entire scene. The pilot and mission specialist roles could in theory be combined and assigned to a single person, but our studies on the safe operations of all modes of UxVs (Murphy and Burke, 2010) as well as the findings of cognitive errors at the recent deployments in Cyprus and Italy (Angermann, Frassl, and Lichtenstern, 2012; Kruijff et al., 2012) argue that a person can just manage to handle one role in the field. Note that this is the case where working with robots in laboratory settings with no pressure, ideal flying conditions, and being well-rested is quite different than the real world.

As noted in Kruijff et al. (2012) and Angermann, Frassl, and Lichtenstern (2012), more people may be involved or needed as well. The mission specialist is often acting on behalf of several different specialists, including the incident commander, structures specialists, hazardous materials experts, and even historians. At the Fukushima Daiichi T-Hawk deployment, there were four people in the lead-shielded control box: a radiation health specialist; the pilot, who because of the design of the UAV OCU had to hold the data collection portion of the mission specialist role; the copilot, who held the diagnostics and flight safety portions of the pilot role; and the flight director, who absorbed most of the conceptual portions of the mission specialist role by making sure that data were collected and detecting unexpected situations that needed to be recorded. At both the Berkman Plaza II (Pratt et al., 2008) and Cypriot naval base (Angermann, Frassl, and Lichtenstern, 2012) flights, an additional person had be added to handle a tether or to maintain a better viewing angle of where the robot was relative to the structure (figure 4.4).

A problem with having a 3:1 ratio is that two of the people are crowded around the pilot's display, which is distracting for the pilot and possibly a mission risk (figure 4.5). Joshua Peschel's doctoral research with responders showed that team performance was enhanced when the mission specialists had their own display to view without looking over the pilot's shoulder (Peschel, 2012a). An iPad that simply mirrors the pilot display was a significant improvement. He went further and created a dedicated display for the mission specialist that removed information of interest only to the pilot and added additional icons and windows with information relevant to a

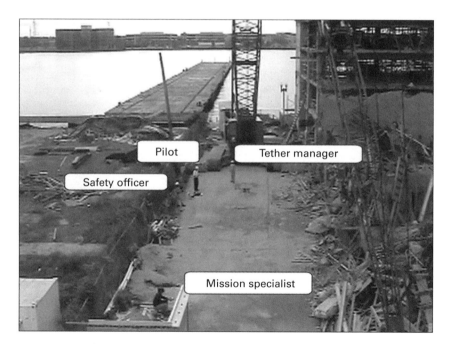

Figure 4.4
A four-person team at the Berkman Plaza II collapse.

CBRNE response and saw an even higher degree of comfort and acceptance from the responders.

4.4.2 Operational Procedures

The operational procedures consist of general steps described below for UAVs operating in local-area environments; the steps will be familiar to civilian pilots. It assumes that the UAV will either be in the line of sight of the operator for safety reasons or in close proximity for interior flights. As a result, the UAV and operator team will use a series of short flights, moving between flights, to capture the faces of the structure or area of interest.

1. *Safety review of site and landing zones selection* This includes checking in with the agency or property owner, determining adequate launch and landing zones, considering the weather conditions, confirming restricted access to bystanders, and projecting a flight path that reduces the possibility of flying near civilians. In post–Hurricane Katrina flights, the operators often had to work from narrow piers to fly on the sides of buildings facing the water, and thus personal flotation devices were required (figure 4.6). In our experience, the selection of safe sites can take from 10 minutes to an hour.
2. *Planning and rehearsal* The pilot leads the checklist, such as making sure the platform and payload are adequately charged and working. The flight director leads the entire team on a brief flight planning session, safety protocol review, and mission rehearsal session (approximately 5 minutes) at the landing zone. Any potential flight problems are discussed, such as flying into the sun or in the interior of structures, and safe zones of operation and backup plans are established. In our experience, this takes only about 10–15 minutes.
3. *Flight* The first flight should be a framing shot or an overview. Subsequent flights will be along each face of the area of interest. Practice has shown that trying to specify the locations for the UAV to take pictures from by GPS is not effective or efficient (Pratt, 2007). Entering GPS coordinates can be time consuming. The native error in GPS and the error induced by loss of GPS strength next to structures make positioning hard. Also, it is difficult for the mission specialist to predict the best locations to examine from the ground. Once in the area, the mission specialist can direct the pilot to good viewpoints. The pilot can compensate for wind by moving the platform to a node and then changing the camera pan or tilt and still produce the same information.
4. *Data review* Once on the ground, the mission specialist should confirm that the necessary data were captured, otherwise the flight should be

Figure 4.5
A pilot flanked by two mission specialists who are giving directions to him while he flies an AirRobot.

immediately reflown. Depending on the mission, the mission specialist might hand off the data or transmit the data to others. In our experience, this takes between 5 and 10 minutes.

4.5   Selection Heuristics

Matching a UAV to a particular mission depends primarily on the intended environment and conditions, the style of operation, the payload, and the ability to record and easily transfer data. Requesting agencies will often ask about the duration of the flight and altitude, but as we've experienced at numerous exercises for local-area operations, it only takes one flight with either a fixed wing or rotary wing for them to realize that they can generally see what they want to see immediately and from lower altitudes than they would expect. For wide-area operations, the duration and altitudes may be different, but they appear largely undocumented.

Below is a summary of heuristics currently used by CRASAR:

Figure 4.6
Pilot and mission specialist standing on a seawall to launch the iSENSYS IP-3 at Biloxi, Mississippi.

- *If the mission is for a wide-area operation, use a fixed-wing UAV, otherwise use a rotary-wing UAV.* There are two exceptions to this rule:

If the launch/landing zone is small, there may be no choice but to use a rotorcraft. For example, at Hurricane Katrina we needed to deploy from a heavily forested, rural small town with trees down everywhere. We had trouble finding enough space safely to hand launch a fixed wing and so used the iSENSYS IP-3 initially because it goes straight up.
If the wind is high, there may be no choice but to use a fixed wing (if the mission permits it).

- *The smaller the UAV, the less likely it is to be able to give a stable image in gusty winds or stay on station.* Many UAV companies show a crisp image taken in some high wind, such as 25 knots, but do not offer any data on how many tries, how long it took, or whether the operator got seasickness from watching the imagery bob up and down until he could get the right image. The options are to get a bigger UAV or try to fly in the mornings before the wind comes up. Even the powerful Honeywell T-Hawk couldn't compete with the afternoon shore breezes and could only fly in the mornings. It should be noted that small and micro UAVs typically do not have hardware image stabilization and neglect the large existing corpus of software methods for image stabilization.

• *If a rotary-wing UAV is appropriate, use one with a "heads-down" interface for long stand-off distances.* The style of operator interface for a UAV falls into three categories: the pilot drives the robot primarily by looking at the robot (heads-up), by looking at an interface (heads-down), or, less commonly, a mixture of both. Heads-up is preferred when the robot needs to fly in cluttered environments because the operator can see the relative position of the robot to the structure. However, at the stand-off distances seen for chemical or nuclear events, heads-up isn't possible, and the team just has to be more careful in clutter.

• *Pick the UAV with the best optics (including image stabilization), ability to pan, tilt, and zoom the camera, and ability to add other sensors such as infrared.* At this point in time, the primary task for a UAV is to provide visual data, so that is crucial. Forward looking infrared red (FLIR) and EO/IR sensors allow responders to see at night and detect people or even the amount of fluid left in a chemical train car; being able to add or swap these sensors to a UAV is desirable.

• *The UAV must be able to record the data, preferably on the OCU, and transfer it.* Many UAVs use a wireless consumer video camera as the payload. The operator sees a low-resolution video from the camera that was intended to serve for the viewfinder, and then when the operator takes a photograph, the high-resolution image is stored on an SD card on the robot. If the UAV crashes, the high-resolution imagery is lost. Also, the video itself is often as important as tiled imagery because it helps provide cognitive context cues, so there should be a way to record it. Along these same lines, the recorded data will most likely be shared with other members of the response. At the very least, there should be a way to transfer the recorded data to a thumb drive, but in general there should be a wireless Internet capability as more and more agencies set up their own LANs.

4.6 Surprises, Gaps, and Open Research Questions

UAVs are the newest of the rescue robot modalities yet are still being teleoperated at disasters. The implementation gaps reflect what response professionals would like to see in a small or micro UAV, but there are surprising gaps in how the UAV community collects and reports data (or rather doesn't). Autonomy for *guarded motion, localization and mapping, human–robot interaction*, and *multiagent coordination* is the major open research question, but the lack of robust simulation for testing and evaluation is important as well.

4.6.1 Surprises

Possibly the biggest surprise about UAVs is that *they were all teleoperated, even the three that were documented as being capable of navigational autonomy and were used in wide-area environments*. Note that the flight stabilization or "neutral joystick" automation stayed on, but the UAVs were actively directed by a pilot rather than given GPS waypoints. Also, it is not known if the UAVs at the Thailand floods or the unsanctioned flyover at Fukushima Daiichi had autonomous capabilities and if they were used. As described earlier in this chapter, UAVs operating close to structures are performing opportunistic movements and may be closer to the structure than GPS error bounds, so it should not be that much of a surprise that teleoperation is used for local-area operations. But the pilots themselves for the Honeywell T-Hawk turned off the autonomous navigation even in transit over distances of up to 1 kilometer from the launch site to the Fukushima Daiichi reactor buildings in order to maintain positive control. That way, they knew that the robot was moving only due to the directives or the wind. This helped them stay prepared in case the UAV failed suddenly due to radiation. The Raven at Hurricane Katrina and the Elbit Skylark at the Haiti earthquake were teleoperated despite wide-area operations because the operators were opportunistically searching and thus adapting the path as this flew toward waypoints. The agencies aren't turning off autonomy, the pilots are; this will be an open research issue discussed later in this section.

Another surprise is that *small UAVs, even ones sold for entertainment and used the same way a hobbyist would, are subject to governmental aviation regulations and liability*. In the United States, the FAA regulations make a distinction between hobbyists flying under 120 meters in unrestricted spaces and a company, university, or agency flying the same UAV under 120 meters in unrestricted spaces. These regulations are an unintended consequence of regulations and definitions made in the 1960s, and the FAA, with increasing pressure from professional societies, is moving to change them. But for now, regulations exist. During a disaster, any size of UAV has to have formal permission, called a **certificate of authorization** (COA), to fly in a particular area. An emergency COA takes about 30 minutes to get. Although annoying to get, an emergency COA for a small or micro UAV actually makes sense. Unlike regular life, manned aircraft do not drop below 400 feet unless they are landing, presumably at an airport. But at a disaster, the Coast Guard or other agencies may be performing tactical rescues where helicopters are flying 10 meters from the ground or water; this is extremely dangerous and difficult. Even a micro UAV could cripple the helicopter or distract the helicopter pilot, so if a helicopter pilot sees any UAV (or other unexpected conditions), the mission is aborted. So the well-intentioned, but cavalier, use of UAVs can actually interfere with lifesaving operations. The emergency COA allows the responders to know who is doing what where, a reasonable request.

UAVs can have tethers, though this isn't recommended. While it only takes about 30 minutes to get an emergency COA, sometimes the answer is "no." CRASAR was denied an emergency COA for the Berkman Plaza II collapse because we intended to get forensic documentation (Pratt et al., 2008). Undeterred, we took advantage of a rule that said a tethered vehicle that stayed under 45 meters did not need a COA, and thus flew the iSENSYS IP-3 with a 150-foot line of sailboard cord. If flying at a disaster is stressful, flying with a tether that can snag and that adds weight is even more so. We achieved the mission objective but do not recommend using a tether in complex environments.

UAVs have been deployed with UGVs and UMVs. While there is no indication that multiple UAVs have been in the air at the same time, a UAV has been used in conjunction with a UMV at Hurricane Wilma (Murphy et al., 2008a) and a UGV at the Tohoku earthquake (Michael et al., 2012) and the Finale Emilia earthquake (Kruijff et al., 2012).

While path-planning and waypoint navigation appears to be inefficient for local-area operations, having a UAV autonomously return to a point of interest and give the same camera view is useful. In our experimentation with CBRNE events, the feedback from the responders is that if they find something of interest, they may want to return and look at that area later. So having the UAV fly directly to that waypoint and place the camera at the appropriate pan and tilt is of great value and now standard on our AirRobot platforms.

UAV missions are short, even for wide-area operations. Flights for local-area operations run between 5 and 12 minutes, driven not by lack of power but rather that people are good at comprehending what they see. Once they see the building, they can quickly discern what they need or take photographs for further reference. While wide-area operations have not frequently appeared in the literature, those missions appear short for the same reason as well. Flights for wide-area operations run between 20 and 35 minutes. The Raven at Hurricane Katrina could have loitered over Pearlington and Bay St. Louis longer, but there was no good reason once the video of the region had been captured. The Fukushima Daiichi missions averaged about 35 minutes and were limited by remaining power.

The default human–robot ratio for a UAV is 3:1, not 1:1. While many manufacturers emphasize that the OCU allows a person to look and joystick at the same time, the human–robot interaction data from our deployments as well as the recent deployments to Fukushima Daiichi, Cyprus, and Italy show that trying to be the pilot and the mission specialist at the same time leads to errors, mission failures, and equipment damage. It is hoped that better combinations of interfaces and autonomous capabilities will reduce the ratio to 2:1, which allows one person additionally to be a safety officer and exploits the buddy system.

4.6.2 Gaps

The gaps in small and micro UAVs for disasters reflect two perspectives. The first is the perspective of the response community on what it wants in a system. The second is gaps in research and development.

Our work with hazmat and transportation specialists and law enforcement experts has identified gaps in UAVs from a user perspective, originally documented in Murphy et al. (2012b) and supplemented by Angermann, Frassl, and Lichtenstern (2012):

• Lack of good optical acuity, as structural specialists need to see cracks and materials and CBRNE responders need to read the labels and placards as well as isolate sources of a leaks. Pan, tilt, zoom with a color camera is a minimum. Note that this feature should be independent of the platform; turning the platform to effect a camera pan can lead to collisions in restricted spaces (Kruijff et al., 2012; Murphy et al., 2012b).
• Lack of image stabilization, which interferes with being able to see details.
• Lack of image enhancement to adapt to changing lighting conditions and see in shadows.
• Inability to review photographs while the vehicle is in flight so that the mission does not have to be repeated.
• Lack of infrared payloads, which are needed for night operations and also to help determine the level of material remaining in vessels.
• Lack of chemical sampling payloads (depending on how the airflow of the UAV impacts sensing) or plug-and-pay payload capability where common chemical or radiological detectors could be added.
• Lack of ability to integrate UAV data with satellite imagery and other geographical sources.

There are also gaps in how data are collected for evaluating and comparing platforms and their capabilities. When one of my doctoral students began examining how 18 different micro UAVs were implementing autonomous takeoff and landing, she discovered that reporting was incomplete and inconsistent (Duncan and Murphy, 2012). For example, only seven of the papers reported the wind conditions for the trials. To understand what UAVs are actually capable of, deployments should capture the number of runs, mission duration, wind conditions, altitudes that the UAV worked at, maximum and minimum altitudes it flew (outside of takeoff and landing), autonomous capabilities, number of humans on the team and their responsibilities, and mission failures.

4.6.3   Open Research Questions

Small and micro UAVs pose at least four categories of open research questions in areas related to increasing the autonomous capabilities of the system: *guarded motion, localization and mapping, human–robot interaction*, and *multiagent coordination*. In addition, testing and evaluation is an open area.

UAVs working in local-area operational environments need **guarded motion**, where the human team members can direct the robot and payload without worry of collision. This ability to drive the sensor and not worry about the robot was noted in Kruijff et al. (2012). The advantage is that it should reduce cognitive load and human error. As noted in Pratt and Murphy (2012), guarded motion has been around for decades. But guarded motion is extremely difficult for UAVs working near structures for two reasons. One is because the UAV may be surrounded by hazards; there may be a structure in front for which the operator has misestimated depth or there may be debris or other equipment on the sides of the UAV or below or behind it. A second reason is that the margin for error is small, of the order 1 meter. One solution might be to mount multiple cameras (rather than heavier range sensors that can work in all lighting conditions) to give a 360-degree view, but visually based range systems have much less accuracy.

Localization and mapping is another autonomous capability that is needed for UAVs working close to structures in GPS-denied environments. Angermann, Frassl, and Lichtenstern (2012) call for more research in **visual navigation**. Visual navigation is essential for obstacle avoidance as small and micro UAVs may not have the payload capacity or energy to support the use of a ranging sensor.

Human–robot interaction (HRI) is an area where advances in autonomy can help reduce the high cognitive load on the pilot and mission specialists. There is a clear need for better interfaces and sharing of control to support creating and maintaining situation awareness. This is a very hard problem that has been encountered by the U.S. Department of Defense in the use of these robots for military operations (Murphy and Shields, 2012); however, the focus has been on HRI for reducing the human–robot ratio for the larger classes of UAVs, not general HRI for small and micro UAVs. For example, there is no real understanding of what the minimum set of autonomous capabilities should be (e.g., autonomous takeoff and landing, waypoint navigation, etc.) or the appropriate control style (heads-up or heads-down) for what missions. One rich area of investigation for small and micro UAVs is to focus on creating better interfaces. One of my former students, Josh Peschel, showed as part of his doctoral research that giving mission specialists their own dedicated display, rather than just duplicating the pilot's display, improved the comfort with and utility of the UAV (Peschel, 2012a).

Autonomy is central to teams of multiple robots. The deployment to the Cypriot naval base suggested that multiple UAVs could simultaneously map a face of the structure, accelerating the reconnaissance process (Angermann, Frassl, and Lichtenstern, 2012). Already, three deployments have shown heterogeneous teams of robots where one was a UAV: Hurricane Wilma (Murphy et al., 2008a), the Tohoku earthquake (Michael et al., 2012), and the Finale Emilia earthquake (Kruijff et al., 2012). However, as we and others have seen, if multiple robots are available at a disaster, they are not necessarily from the same agency or group, and thus coordination is difficult (Kawatsuma, Fukushima, and Okada, 2012; Murphy et al., 2012b). A software middleware infrastructure may be the answer to coordination, but it would take a tremendous investment and standardization.

Autonomy and HRI will continue to gain importance as multiple robots become commonplace and as network connectivity grows allowing many distant users to access the robots independently in real time. I have dubbed this the **100:100 challenge** (Murphy, 2011a), where the human–robot ratio will no longer be thought of in terms of how to reduce the ratio from 3:1 to 2:1 or make swarms with a 1:100 ratio but rather how can many users access how many UAVs.

4.7   Summary

Small and micro UAVs are highly portable, organic assets for responders to use on demand, though in accordance with applicable regulations. They have been at 11 deployments to hurricanes, building collapses, floods, and earthquakes, where they have successfully helped specialists understand the situations and also allowed officials to prevent further flooding in Thailand. The use of UAVs is increasing, possibly in part due to the rapid technological advances being made in flight controls as well as their unique capabilities. Small and micro UAVs can provide views from altitudes and angles not covered by manned vehicles or satellites and can provide imagery on demand versus waiting for manned assets, and they are more cost-effective than manned assets.

The difference between the small and micro sizes of platforms is unclear in the community, and a more useful way of thinking about small UAVs for tactical disaster operators is to consider whether they are fixed wing, like a plane, or rotorcraft, like a helicopter, and how they would be used for an operational environment. Rotorcraft UAVs have been primarily used for local-area operational environments, where they may hover as close as 2–5 meters from structures or areas of interest in order to perform structural inspection. Some speculative work is exploring sUAVs flying into buildings from the outside, though this has only been tried once at damaged buildings at Hurricane Katrina. Fixed-wing UAVs have been used less frequently and only for reconnaissance and mapping tasks in wide-area operational environments, though this may be an artifact of the lack of UAV availability. The use of UAVs for interior structural inspection of large churches at the Christchurch earthquake and Finale Emilia earthquake highlights the third operational environment: interiors. As UAVs become more commonplace and less-skilled users want to fly them, the style of control (teleoperated, delegated, and guarded motion) and the pilot primary viewpoint (heads-up or heads-down) may become as important as the choice of platform design.

In practice, UAVs lead to high cognitive workload and a susceptibility to human error. The concept of operations that we created in the aftermath of Hurricane Katrina has been adopted and used successfully at five subsequent deployments by CRASAR and other groups. The resulting recommendations are to

• Assume a 3:1 human to robot ratio where each person is assigned a specific role (pilot, mission specialist, safety officer). Splitting roles between people may be necessary depending on the type of OCU and UAV.
• Follow a straightforward operational procedure that begins with a safety review of the site and selection of the landing zones; planning and rehearsal immediately before each flight; accommodations for the mission specialist to refine the mission objectives during the flight once he or she can actually see the situation; and data review on the ground.
• Give the mission specialist a display to prevent the specialist from crowding the pilot, either a mirror of the pilot's display or a display designed for the specialist's needs (preferred).

The response community appears eager to adopt UAVs but has a long list of implementation gaps. The most serious gap is the general lack of good optical acuity, image stabilization, and image enhancement. The mission for a UAV is to be the remote eyes of the responders, so image quality is essential. The ability to add infrared or chemical payloads is important as well. As a practical matter, the camera systems on UAVs should have an independent pan and tilt rather than being fixed and relying on the UAV to turn to provide a view.

UAVs have been teleoperated in conjunction with onboard flight stability software where the UAV maintains the last position if the pilot lets go of the controller paddles. However, the lack of autonomy has a high cost in cognitive overload and risk of the vehicle. Open research issues center upon increasing the autonomous capabilities of the system: guarded motion, localization and mapping, human–robot interaction, and multiagent coordination. In addition, robust simulators are needed for testing and evaluation.

Basic research projects are looking at concepts that could eventually be of great value to emergency management. The abilities to avoid trees, power lines, and manned vehicles, to maintain position relative to an object of interest, to create three-dimensional maps, and to fly indoors are popular research topics. One approach is to assume ranging sensors will eventually become small and light enough to fit on small and micro UAVs. Another promising approach is to rely strictly on vision, as a UAV will always have a camera. UAV swarms or a team of UAVs is another topic with an active community, and the use of UAVs to carry and drop off sensors and manipulate the environment is emerging as well. Another line of investigation is the use of UAVs to help with evacuation after an incident.

5 Unmanned Marine Vehicles

Chapter 2 described how unmanned marine vehicles are a growing segment of rescue robotics. Of the 34 incidents where robots have been used, unmanned marine vehicles (UMVs) have been present at seven, putting UMVs in last place for frequency of deployment. But in terms of the number of robots deployed, there were at least 31 UMVs used for those seven incidents, making a close second in total number of robots used to date. This chapter focuses on UMVs and their increasing role in disaster response and recovery.

By the end of this chapter, the reader should be able to:

• Describe the three types of UMVs: **autonomous underwater vehicles**, **remotely operated vehicles**, and **unmanned surface vehicles**.
• Describe the common missions and tasks for UMVs.
• Demonstrate familiarity with where UMVs have been used.
• Demonstrate familiarity with the performance of UMVs and their common modes of failures.
• Match a UMV(s) with the constraints imposed by the environment and the necessary tasks for the mission using the selection heuristics.
• Show familiarity with the gaps in the technology and the open research questions.

The chapter begins by describing the three types of UMVs, followed by description of the environments the UMVs are expected to work in. A case study of the use of UMVs to inspect the damaged Rollover Pass Bridge after Hurricane Ike illustrates some of the challenges of littoral environments after a coastal event. Understanding the capabilities of the robots and the constraints of the environment aids in comprehending the range of missions and tasks for UMVs. The chapter then discusses where UMVs have been used, their performance, and their failures. A case study from the Tohoku tsunami provides hard data on how effective UMVs can be, despite environmental constraints. The discussions of robots, environments, missions, and history of use lead to a set of selection heuristics for matching a UMV to a particular disaster. The discussions also highlight surprises about UMVs and raise open research questions for future work.

5.1   Types of UMVs

UMVs fall into three broad categories: **autonomous underwater vehicles (AUVs), remotely operated vehicles (ROVs),** and **unmanned surface vehicles (USVs).** As AUVs and ROVs typically operate fully submerged, they may also be referred to as **unmanned underwater vehicles (UUVs)** (figure 5.1). In practice, the distinguishing attributes between the categories for rescue robotics are the following: whether the vehicle operates in a remote presence or taskable agent mode; the launch and recovery constraints; the sensor payload; and the power demands.

Figure 5.1
Examples of UMVs: (top left) a YSI Oceanmapper AUV, (top right) a SeaBotix SARbot ROV, and (bottom) an AEOS Sea-RAI USV.

5.1.1   Autonomous Underwater Vehicles

AUVs swim without use of any tethers, thus earning the title of "autono-mous," which implies the greater artificial intelligence abilities that they typically possess. In most cases, AUVs are simple automatons, diving below the draft of surface watercraft and using dead reckoning to follow preprogrammed routes through volumes of water where no obstacles are expected—a marine variation of the "big sky, small plane" assumption in unmanned aerial vehicles (UAVs). AUVs typically carry a side-scan sonar to map the region as they travel and one or more sampling sensors, such as for water quality. Power is derived from onboard batteries or solar cells, which can limit duration, though deepwater systems can make use of ther-moclines and waves to minimize onboard power needs. An AUV can be launched from the shore or a boat and may require a chase boat to retrieve it in case of a problem.

AUVs offer many advantages but present some disadvantages for rescue and recovery. Because they operate submerged, they do not interfere with or restrict marine traffic so they can be used in littoral regions without restrict-ing transportation of responders into an area or shipping of relief supplies. They can cover large areas such as a bay very quickly (i.e., in hours), and a new style, gliders, can conduct missions lasting weeks or months. AUVs derived from the Oceanmapper design are small and can be transported in a car or dinghy and carried by one or two people. With respect to disad-vantages, AUVs are taskable agents, not remote presence devices, as they do not communicate with operators except when they surface to use GPS and may not have sufficient bandwidth and power to transmit data in real time. They depend on the accuracy of tide and depth charts for operations in littoral regions and generally do not have highly accurate localization for survey of bridges. An AUV may not be powerful enough to overcome cur-rents and tides and without accurate localization may not know it is going backwards.

5.1.2   Remotely Operated Vehicles

ROVs are underwater robots controlled by a tether. The movie *Titanic* opens in the present day with two actors using an ROV to explore the wreck. ROVs are used for operations with underwater structures, such as oil/gas, or for aquaculture. ROVs are classified by size, weight, ability, or power (Wikipedia, 2012). Smaller ROVs such as the *mini or micro* classes can be transported by a car and one or two people and launched from shore, a dinghy, or off a bridge or seawall. They are designed to go into areas where a diver cannot fit. All known ROVs carry a camera and usually some type of imaging sonar, or acoustic imager, that can provide real-time sonar images. Power is provided through the tether, and the energy demands for the ROV motors, active sonar, and operator control unit may require a generator. Because ROVs are tethered, navigation favors a spherical coordinate system representation.

ROVs pose a vexing combination of advantages and disadvantages. ROVs are portable and provide remote presence. But, as with an unmanned ground vehicle (UGV), the tether is both an advantage and a disadvantage. The tether provides power, permits real-time remote presence, and serves as a safety line. However, it can tangle around obstacles in the water and is subject to drift by currents, both increasing the possibility of tangling and pulling the ROV off course. The tether also limits the area of operations. An ROV may be more powerful and better able to overcome current than an AUV, but this requires power. While some ROVs may tap a car or marine battery, others require a generator, which increases logistics demands. An ROV can be easy to deploy, but the operator needs a nearby area in which to set up the operator control unit (OCU) and to work from; there is something counterintuitive about standing on a bridge in order to lower an ROV to determine if the bridge is about to collapse. Because mini ROVs work within only a few hundred meters of the work area, a second sonar system can be placed at the launch point to detect the robot's location.

5.1.3 Unmanned Surface Vehicles

USVs are robot boats. There is no widely accepted classification of USVs, with the U.S. Navy defining USVs either by the length of the waterline or by displacement. USVs for search and rescue are typically the smaller members of the X-class (waterline length of less than 3 meters). They are too small to carry a person but large enough to carry a significant computing and sensing payload. USVs take one of four form factors: air-breathing submersible, jet-ski based, rigid inflatable boat based, or kayak or pontoon based. With the exception of air-breathing submersibles, USVs generally have a shallow draft, and those such as the AEOS series can operate in 15.25 cm of water.

Commercial USVs are used primarily for marine biology but are also being explored for homeland security missions such as port security and rapid interception of watercraft. One USV, Emergency Integrated Lifesaving Lanyard (Emily), is being sold for use by lifeguards and has reportedly saved a father and son trapped in a riptide (Smith, 2012). All known USVs carry some type of payload for sensing above the waterline, such as a camera. Jet-ski—based platforms do not carry underwater sensors, limiting their utility for rescue and recovery; they are intended to cover large distances to intercept intruders. Power is onboard, but marine batteries appear sufficient for at least 8 hours of operation. Launch and recovery of a USV depend on its size, but the inflatables and kayak/pontoon platforms are designed to be small enough and light enough to be manually carried to the shore or lowered into the water from a structure without access to a boat ramp.

USVs present four major advantages but also three disadvantages. They are free swimming and avoid the limitations of a tether. They are above the water, so can receive GPS and maintain a wireless network; this means they can operate either with full autonomous or with remote presence and transmit data in real time. They can carry a larger sensor payload and sense both above and below the waterline. If the payload includes an imaging sonar, the USV can see through turbidity. However, USVs are generally larger and harder to transport. In addition, the operators are responsible for meeting *marine collision regulations* (International Regulations for Prevention of Collisions at Sea; often abbreviated as **COLREGS**). A professional society for unmanned systems, the Association for Unmanned Systems International, is leading the effort to update the collision regulations for USVs (and for AUVs when they surface to transmit data, to recharge, or to be recovered), particularly to clarify if USVs can be treated as a "vessel restricted in her ability to maneuver" as opposed to having to provide obstacle-avoidance capabilities. A more subtle disadvantage is that GPS and wireless signals are disrupted by urban littoral structures such as bridges as documented by Murphy et al. (2011b), which affects navigation and control.

5.2   Environments

The marine environment for robots can be classified either as **littoral** or **deep water**. Most UMV rescue and recovery operations have been in littoral regions, with the exceptions being the Deepwater Horizon (2010), missing balloonists (2010), and *Costa Concordia* (2012) events. There is some debate within the IEEE Safety, Security, and Rescue Robotics Technical Committee (the professional entity devoted to these efforts) as to whether deepwater operations should be treated differently than nominal search, rescue, and recovery operations. This section will concentrate on operations in the littoral environment, first defining what is meant by littoral regions, the access to littoral regions after a disaster, the condition of the water, and the availability of GPS.

5.2.1   Littoral Regions

Response and recovery operations in littoral regions present unique constraints on robot deployment. There are multiple definitions of *littoral region*, all of which essentially converge on a littoral region being the shallow areas extending from the shore. Ports, piers, bridges, marinas, and inlets protected by breakwaters are examples of *urban littoral regions*, where man-made structures have been inserted into a body of water. Bays, lakes, and rivers also fall into the category of littoral region.

5.2.2 Access
Access to littoral regions is usually limited, especially as boat ramps and boats are damaged. In general, the UMV team has to be able to transport the marine vehicle and supporting gear through damaged areas to the littoral region of interest, manually carry and launch the UMV, and power the robots, as marinas, gas stations, and electricity may not be available or functioning. The team cannot rely on a chase boat to deploy or to recover an untethered UMV. The access conditions during the response phase favor small ROVs that can be deployed from the shore, the structure, or a small boat (figure 5.2).

Coastal boat ramps are usually destroyed, and a working boat ramp further inland may be kilometers away, requiring travel through unmapped shallows as the depth charts are now uncertain. Boats may not be available

Figure 5.2
Example from the 2011 Tohoku tsunami response of the challenging UMV launch, operations, and recovery access conditions.

on site as local boats are often damaged or destroyed; of the 1,000 registered fishing boats at Minamisanriku, Japan, only 56 survived in working order (Leitsinger, 2011). Authorities may bring in boats, but those will generally be committed to transporting high-priority rescue workers and supplies. Even if boats are available, conditions may not permit a boat on a trailer being driven to the water, thus favoring small inflatables, which are at a premium after a water-based disaster.

Fortunately, the majority of urban structures are accessible from either the structure itself or from the shore, which suggests operating distances of the order 300–500 meters. This is generally feasible for deploying tethered ROVs and for wireless line-of-sight operations.

5.2.3 Water Conditions

The water conditions in urban littoral regions are very demanding in terms of navigability, visibility, and currents. These conditions drive the choice of UMV and also favor systems that have both sonar and video sensors.

The disaster may have changed the shoreline and has most likely deposited debris and changed the underwater topology, rendering depth charts useless. New satellite or geo-registered overflight data may not be available for more than a week, and it may be a year or more before depth charts are updated. This may prevent AUVs from operating below the water as they typically have no obstacle-avoidance capability and can be difficult to retrieve.

Visibility can be a problem throughout rescue and recovery and favors the use of sonars. Initially, the water in normally clear regions may be extremely turbid (cloudy) due to suspended sediment and runoff. As the time goes on, underwater vegetation may quickly sprout and cover up objects. Sonars penetrate through turbidity, so often teams will use the sonar on the UMV to locate and navigate to an object of interest, then use a video camera to determine the details. UMVs with only underwater video cameras may not be useful.

Currents and the external influences of wind, waves, and tides can interfere with the control of the UMV. Currents affect AUVs more than USVs, which can use GPS to maintain position, and ROVs, which can be retrieved by their tether. AUVs and ROVs are less susceptible to wave action. All types of UMVs working in shallow areas are vulnerable to depth changes due to tides.

5.2.4 GPS-Denied Areas

Littoral regions present GPS or localization challenges for UMVs. If the AUV is submerged, it has no GPS signal and must rely on inertial sensors and Doppler velocity sensors to maintain course. ROVs can use a transitive sonar scheme, where a sonar system can be put in the water at the launch point. The GPS location of the launch point is known, and the approximate location of the ROV relative to the launch point is known from the sonar, thus the ROV location in GPS coordinates can be inferred. Unfortunately, this transitive sonar approach has not been accurate enough for control. USVs in theory should have GPS signal in littoral environments, but in practice GPS dropout may occur due to "shadows" from urban structures such as bridges (Murphy et al., 2011b).

5.2.5   Case Study on UMV Environments: Rollover Pass Bridge

Our experiences inspecting the partially collapsed Rollover Pass Bridge in Texas for the Texas Department of Transportation after Hurricane Ike (2008) shows some of the problems introduced by the marine environment. Two months after the hurricane, CRASAR brought an AEOS Sea-RAI USV and a Video Ray ROV and were joined by YSI who brought their Echomapper AUV. The Rollover Pass Bridge connected two parts of the Bolivar Peninsula along a beach. It was especially challenging for manual divers as the currents were so swift that it was safe to dive only for 15 minutes at slack tide. This also meant that the currents might be too much for the USV or the AUV, so a chase boat would be helpful, especially considering the USV cost just under $500,000. However, there was no guarantee of a working boat ramp nearby.

To solve the chase boat problem, CRASAR asked to borrow a small inflatable boat from the Texas Engineering Extension Service (TEEX) at Texas A&M University. However, TEEX knew from conducting rescue work at the Bolivar Peninsula that there was an intact boat ramp about 1.5 kilometers from where we would be and that we could drive a trailer into that area. Therefore, one of the team members could launch a much more powerful boat that could track a rogue UMV farther off shore and in worse conditions.

However, the hurricane had significantly rearranged the topology of the inlet, and on-site conditions were foggy. As we set up the USV, the boat was launched but ran aground and had to wait for the high tide to refloat. Meanwhile, as there was no cell phone coverage, we had no idea what was wrong. Eventually, we went ahead and launched the USV from a sand spit with a 150-foot-long cord used for sailboarding as a safety line until we were sure the motors could handle the current. Figure 5.3 is a photograph of me walking the USV as if it were a dog about 200 meters from the sand spit up to the bridge. We took off the safety line after about an hour, having gained confidence in the UMV's motors.

Figure 5.3
The Sea-RAI with a temporary safety line at the channel for the Rollover Pass Bridge.

5.3 Missions and Tasks

UMVs have been used for seven of the rescue robot missions identified in chapter 1. It should be noted that a mission for a UMV operating in an urban littoral region may be a combination of missions. Consider that the region of interest in inspecting a bridge spanning a river is both the substructure of the bridge and an area upstream that may contain submerged debris that could sweep down and inflict damage. Likewise, having a clear channel around a pier for access is as important as the structural integrity of the pier itself. Understanding the seven missions for UMVs (*structural inspection; estimation of debris volume and type; victim recovery; forensics; environmental remediation; search, reconnaissance, and mapping;* and *direct intervention*) leads to understanding the types of tasks that a UMV should be competent at. The types of tasks or competencies fall into three categories: *mobility, manipulation,* and *perception*.

The seven missions UMVs have been used for are as follows:

• *Structural inspection*   As seen at Hurricane Wilma (Murphy et al., 2011b), the I-35 bridge collapse in Minnesota (FBI, 2007; Goldwert, 2007), and at

Hurricane Ike (Murphy et al., 2011b), UMVs can be used to inspect littoral urban structures such as bridges, seawalls, piers, and docks. The UMVs can investigate direct damage to the structure but also map out upstream debris fields that may cause damage in the future.

• *Estimation of debris volume and types*   Determining the amount of rubble is necessary for both the allocation of necessary resources to remove the rubble and for planning on how to dispose of it. The use of UMVs at the Tohoku tsunami recovery operations (2011) relied on the UMVs to find the debris but humans to estimate volume.

• *Victim recovery*   UMVs searched for the bodies of victims at the Minnesota I-35 bridge collapse (2007) and during the Tohoku tsunami response (2011). The capsize of the *Costa Concordia* (2012) and the missing balloonists incident (2010) occurred in deeper waters, with ROVs performing a particularly difficult search through the interior of the wreckage of the *Costa Concordia*.

• *Forensics*   In cases where there is evidence of terrorism or negligence, UMVs may provide forensic evidence. For example in the Minnesota I-35 bridge collapse, civil engineers needed to understand why the bridge had fallen. This in turn meant the engineers had to document the underwater portions of the substructure and where the bridge elements had fallen before they were swept out of place by the strong currents.

• *Environmental remediation*   Understanding what is polluting the water or other situations that can affect the quality of the biosphere and livelihoods is an important element of disaster mitigation, either during the response or the recovery phase. A UAV used at Hurricane Wilma saw an unexplained plume in the water near Marco Island; a UMV could have investigated and sampled it.

• *Search, reconnaissance, and mapping*   Searching, reconnoitering, and mapping missions are often central to other missions; for example, mapping an upstream debris field is part of assessing what has to be done to allow a bridge to be reopened. Another example is investigating whether a channel can be reopened for navigation. The need to search large areas was a common thread in Hurricane Ike (2008), the Haiti earthquake (2010), the Minnesota I-35 bridge collapse (2010), and the Tohoku tsunami (2011).

• *Direct intervention*   UMVs can go beyond passive examination of an underwater scene: As seen in the Deepwater Horizon event (2010), ROVs were central to both determining the situation and to manipulating the blowout preventer and installing the remediation cap. Direct intervention is also needed for extracting a person from a car or physically grabbing and bringing back an object of interest.

These missions listed above suggest that a competent UMV will be capable of performing the following tasks:

• **Mobility, including localization** To search, reconnoiter, and map underwater areas, the robot has to be able to move in littoral regions, to know where it is, and to know where it has been with certainty. The different types of UMVs provide mobility, but often do not provide localization. Our work at Hurricane Wilma (Murphy et al., 2008a) showed that the operator lost track of which piling was being examined, as the GPS error was of the order the piling spacing. Reducing localization and mapping uncertainty is essential for accuracy but also human confidence. Our work at the Tohoku tsunami recovery showed that confidence in the coverage of the UMV was critical for acceptance by users (Murphy et al., 2012a); they wanted proof that the robots would cover the entire area with sufficient sensing to be sure to find objects of interest.

• **Manipulation** Manipulation tasks fall into two categories: *investigatory* and *intervention*. Manipulation can enable investigatory actions, also known as *active perception,* where the robot prods or pokes at an object to help disambiguate what it is. An example is shown in figure 5.4 where the end effector on the SeaBotix SARbot was used at Rikuzentakata, Japan, to probe a possible hand and determined it was a rubber glove. Manipulation can also be used for intervention, especially to break car windows, cut seat belts, and grasp objects or victims. Figure 5.5 shows the SARbot retrieving a child's backpack.

• **Cooperative perception** Although the unpredictable nature of disasters precludes fully autonomous object recognition, intelligent algorithms are needed to assist humans in perception. As witnessed at the Tohoku tsunami, the image enhancement software provided by the LYYN company significantly improved video visibility. Sonar tiling at the Rollover Pass Bridge gave an overview of the debris field that was far easier to understand than the view from video where the USV was moving about. To support human interpretation of the sonar and video data, further processing is needed. Algorithms that help detect possible objects of interest, such as bodies, or indicators of structural problems, such as scour under bridge footings, can direct attention to the data and make sure the human has noticed this possible feature.

5.4 Where UMVs Have Been Used

Table 5.1 summarizes the seven known UMV deployments. Each row lists the robots used (if known) for each event by type and the missions. If more

Figure 5.4
View of what appeared to be a human hand but through investigatory manipulation
was determined to be a glove.

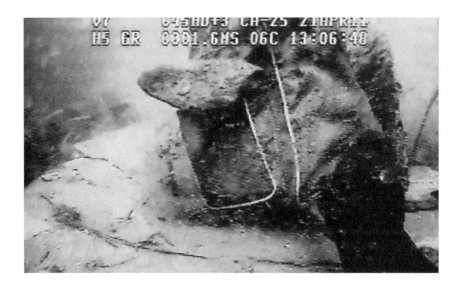

Figure 5.5
SARbot gripper manipulating a child's backpack.

than one type of UMV was used at an event or for different missions, the types are listed. For example, the mini Access AC-ROV used at the Tohoku tsunami did not have sonar and thus could not be used for the same missions as the larger ROVs.

The table shows a preference to date for ROVs. The majority (6) of the incidents used ROVs, while USVs and UUVs were used in only two events. Structural inspection is the prevalent mission, with four events using at least one UMV to examine critical infrastructure. General search, estimating the debris, and searching for victims are the next most popular tasks. Environmental remediation and forensics are the other missions that have been reported.

Note that the historical use of ROVs does not preclude a shift to USVs or UUVs for future disasters or new applications, but they do suggest what is valuable to users: *portability, flexibility,* and *low risk.* The tether provides real-time perception and control as well as security that the vehicle can be recovered. While the tether limits the vehicle's distance of operations, most littoral structures are close to land or accessible from shore. Autonomy is either not accurate enough for littoral missions or is perceived as risking the vehicle.

5.4.1 Performance

The performance of a UMV can be measured in terms of **mission success, findings, time in water** (aka mission duration), **setup time,** and **errors.** *Mission success* is generally coverage of the volume of interest to a subjective level of confidence; localization may not be sufficient to guarantee that the UMV did cover everything. The coverage may produce *positive findings,* such as found structural damage or a sunken boat, or *negative findings* that there was nothing of note in the volume. Metrics of usefulness are the *time in the water* to accomplish the mission and *setup time.* Shorter times in the water indicate effectiveness over a manual diver, and shorter setup times suggest that the system is portable and more attractive to responders. The number and type of *errors* that occurred is also important to capture; errors are discussed separately in the next subsection.

UMVs have shown *mission success* and produced important *findings,* though as seen in section 5.3, these missions tend not to be newsworthy. Ports and shipping channels in Japan and Haiti were reopened using ROVs, allowing relief supplies to enter devastated areas and economic activities to resume. While only two bodies have been recovered using UMVs, the fact that they were used in Italy and Japan to search for victims is noteworthy because the ROVs ruled out areas for manual divers and sped up the

Table 5.1
Known UMV deployments

Year	Event	Robot	Type			Mission						
			USV	ROV	UUV	Structural Inspection	Estimation of Debris	Victim Recovery	Forensics	Environmental Remediation	Search	Direct Intervention
2005	Hurricane Wilma (USA)*	AEOS-1	X			X						
2007	I-35 Minnesota bridge collapse (USA)	Video Ray plus another		X				X	X			
2008	Hurricane Ike (USA)	AEOS Sea-RAI	X			X	X					
		Video Ray		X		X						
		YSI Echomapper			X		X					
2010	Haiti earthquake (Haiti)	SeaBotix LBV		X		X						

Table 5.1 (continued)

Year	Event	Robot	Type			Mission						
			USV	ROV	UUV	Structural Inspection	Estimation of Debris	Victim Recovery	Forensics	Environmental Remediation	Search	Direct Intervention
2010	Deepwater Horizon (USA)	Schilling Ultra Heavy Duty, Oceaneering Maxximum and Millenium ROVs		X							X	X
2010	Missing balloonists (Italy)	Unknown ROV		X							X	
2011	Tohoku tsunami (Japan)	SeaBotix SARbot		X		X	X	X		X		
		Seamor		X		X	X	X				
		AC-ROV		X				X				
		Hirose custom		X				X				
		Ura unknown		X				X				
		YSI Echomapper			X		X	X				

search. UMVs have provided valuable forensic information at the Minnesota bridge collapse and are being considered by state departments of transportation for bridge inspections. ROVs in Japan helped remediate fishing beds and accelerate the restarting of the fishing economy north of Sendai after the Tohoku tsunami.

The *time in water* is not known for all deployments, but in general the missions are relatively fast. As with UGVs, once the operator could perceive the situation or object, the actual classification or decision was immediate. The USV deployment to Hurricane Wilma was only 1 day, and the dock and pier were examined in about 1 hour of actual operation time, with the rest of the time spent on experimentation and repeating the runs. The USV deployment to the Rollover Pass Bridge after Hurricane Ike was similar in duration, with a thorough examination for scour on all the pilings taking about 2 hours. As with Hurricane Wilma, the bulk of the time was spent on taking advantage of the site for experimentation. ROV data are also scant, but the most recent data from the Tohoku tsunami showed the ROVs were able to cover approximately 2,400 square meters in 9.25 hours of time in the water when deployed from the shore and 80,000 square meters in 6.2 hours when deployed from a boat. There is no data on AUV time in water at disasters.

The *setup time* is also not known for all deployments, but for the CRASAR events the shortest time was 10 minutes (SeaBotix SARbot) and the longest about 1.5 hours (AEOS-1 USV). The USVs took the longest, of the order 1–1.5 hours to go from the case to the water and passing the systems check for sensors and network communications. This is because USVs had to be taken apart for shipping and reassembled on-site. AUVs were faster, about 40 minutes to unpack and to check out the sensors. The ROVs took between 10 and 40 minutes for setup depending on the model, as they typically require no assembly, just attachment to the operator control unit.

5.4.2 Mission Failures

Returning to table 5.1, of the seven deployments, mission failures were encountered six times. These failures fell into three categories from the taxonomy in chapter 1: **externally induced**, where a robot could not be used at all due to the environment (2 failures), **physical failure**, where a robot was damaged in shipping (1 failure), or **human error** that led to a mission termination (3 failures). If a UMV is matched to environment and mission, it physically performs well. Human error is the problem.

Externally induced failures stem from an event or environmental factor. In general, UMVs are well designed for physical survival in the environment;

indeed, a major aspect of the cost is waterproofing and resistance to corrosion. As noted earlier, there is always the problem of whether the UMV can operate in the current and clutter of the environment. But beyond being wet, the current and turbidity in marine environments may lead to failure of the UMV to accomplish its mission. For example, the ROV at the I-35 bridge collapse in Minnesota was believed to have encountered difficulties with both the current and turbidity of the water (FBI, 2007; Goldwert, 2007). The nature of urban littoral regions also creates environmental factors that can lead to failure. The UUV at the Hurricane Ike inspection of the Rollover Pass Bridge experienced localization errors due to GPS error and loss of signal (Murphy et al., 2011b).

The two known instances of *physical failure*, where a key component of the UMV simply did not work, were related to shipping. The two failures were at the Tohoku tsunami deployments. The operator control unit for the Seamor ROV did not work upon unpacking but was repaired in the field. The YSI Echomapper also did not work when unpacked but could not be repaired. This "dead on arrival" phenomenon illustrates that successful robotics isn't just about designing a robot that can work in harsh conditions but also about rugged construction and packaging and ease of repair or replacement.

Three of the deployments experienced *human error*, though in fairness to the operators, the expectations of what the human could do appeared to be too high. At both the Hurricane Ike and the Tohoku tsunami deployments, the tether for an ROV became wrapped around underwater structures. In these cases, the lack of visibility combined with the lack of general situation awareness of the relationship of the robot, the tether, and the structure led the operator to tangle the tether more. In both cases, the missions had to be terminated and extensive effort made to recover the platform. The Deepwater Horizon deployment saw two collisions of the ROVs with the structure (Chazan, 2010; Newman, 2010) and possibly collision of one ROV with other ROVs (Bob99Z, 2010) that temporarily halted the mitigation effort. Two of the collisions terminated the mission and forced the operators to redo the work of initially inserting the tube into the blowout preventer and later to remove the lower marine riser package (LMRP) cap. In each of these cases, the operator was expected to infer and maintain a complex mental model of where the tether or robot was over time using indirect sensing.

Solutions to these environmental, physical, and human failures depend on better sensors and sensing. UMVs need better sensors and image processing to operate in turbidity. Proprioceptive sensing to estimate the position of the tether is needed; companies such as Video Ray already sell coarse systems. Better localization and general visualization of where the UMVs are relative to structures and other UMVs is needed; companies such as Sea-Botix deploy an external sonar to generate a rough estimate of localization, but this does not appear to be commonplace.

5.4.3 Case Study: Tohoku Tsunami

The use of UMVs by the International Rescue System Institute (IRS) and CRASAR for recovery operations at the Tohoku tsunami serve as a case study for two aspects: (1) how a disaster affects the choice of UMVs and constrains the use of multiple UMVs; (2) the significant speed and area of coverage that UMVs can provide. More details about the deployments can be found in Murphy et al. (2012a).

The Tohoku tsunami on March 11, 2011, destroyed more than 700 kilometers of shoreline, most of the ports and bridges, and rearranged shipping channels. Given the extent of the coastal damage, it seemed reasonable that unmanned marine robots would be able to accelerate response and recovery efforts. On March 13, the Port of Hachinohe requested CRASAR and the IRS to provide UMVs to help reopen the port. Hachinohe is a major port at the northern end of the Japanese main island and a primary route for relief supplies. Because of travel restrictions imposed by the rapidly worsening Fukushima Daiichi nuclear accident, CRASAR could not enter Japan for several weeks; in the meantime, Hachinohe had been reopened. Eventually, CRASAR was able to join IRS in the field to meet the requests of the municipalities of Minamisanriku and Rikuzentakata from April 18 to 24, 2011, with four ROVs working from shore, where they reopened ports and searched for submerged victims. CRASAR returned with IRS from October 18 to 28, 2011, with an ROV and AUV and cleared fishing beds from debris and pollution while working from a boat. The October deployment covered an area 40 times larger than in April in less time, highlighting the limitations from working from shore after a disaster.

In both the April and October deployments, the mission requirements and the logistics constrained the choice of robots. The requested mission in April was port reopening, which has two key characteristics. First, the robots would be working in a localized area versus an entire bay. Second, the robots would be controlled from land to cover those areas as boats were unavailable. As with coastal disasters, boats had been destroyed by the tsunami or were already committed to more critical operations. The logistics of transporting an inflatable boat and motor on commercial airlines and then into the ravaged area were unrealistic. Therefore, ROVs that could be deployed from shore or structures and had tethers to ensure that the platforms would not be lost were a good choice. The amount of floating debris ruled out an AUV. It might encounter debris, and a malfunction would require a boat to recover the device. A USV was attractive as being able to work in both shallow water and over large distances, but USVs have a larger footprint than small ROVs and AUVs, and there were no trucks available to carry a USV to the site, only minivans.

The team selected four ROVs: an Access AC-ROV (CRASAR), an LBV-300 and a SARbot (both courtesy of SeaBotix) with video image enhancement (courtesy of LYYN), and a Seamor with a DIDSON sonar payload (courtesy of the University of South Florida). The SeaBotix LBV-300 and SARbot are similar, but the SARbot has reduced functionality in order to support rapid deployment. The DIDSON imaging sonar carried by the Seamor had a higher resolution than that carried by either of the SeaBotix vehicles, but the ROV itself was less agile. The Access AC-ROV was essentially an underwater camcorder with thrusters; it did not carry a sonar but could get into small locations. All robots were teleoperated, as is standard with ROVs. Figure 5.6 shows the three robots; the LBV-300 was not used.

Figure 5.6
The three robots used by the IRS—CRASAR team in the Tohoku tsunami April 2011 deployment. From upper left, clockwise: SeaBotix SARbot, Seamor, and Access AC-ROV.

Figure 5.7
UMVs used by the IRS—CRASAR team at the Tohoku tsunami October 2011 deployment.

The requested mission in October was to help the local fishing collectives map the remaining debris in the prime fishing and aquaculture areas, particularly looking for cars and boats leaking oil and gas and for debris that would snag and tear fishing nets. This mission required working over large areas of the bay, and boats would be available. The size of the area to be searched and the removal or settling of floating debris suggested that an AUV would be valuable in rapidly mapping out a large area. Then an ROV could be used to go to suspected debris and confirm if it was an outcropping of rock or debris to be removed. The team selected a YSI Ecomapper AUV (courtesy of Oceanmapper and YSI/Nanotech Japan) and a SeaBotix SARbot (courtesy of SeaBotix), shown in figure 5.7.

Despite having four robots on the April deployment, the fieldwork rarely used multiple robots. One reason was due to personnel safety: Even though we sufficiently trained personnel and cars to divide into two separate teams, the group stayed together because of the lack of cell phone coverage to ensure communication, coordination, and safety in case of another tsunami. Another reason was that it was often difficult to use two robots at the same time because of the *lack of workspace* and *sensor interference*. Figure 5.8 shows the typical workspace available. There was rarely a 1-meter radius of space available safely to set up the gear and run the operator control unit from. Marinas and ports provided a larger workspace where two ROVs could be used at the same time, but the robots had to have sonar operating at different frequencies or be sufficiently separated to prevent the active sonars from interfering with each other.

The October deployment was specifically designed to be a multirobot deployment but degenerated to a single-robot deployment because of an equipment failure. The intent was for the AUV autonomously to map out

Figure 5.8
Examples of constrained workspaces from the Tohoku tsunami April 2011 deployment.

large areas of the bay, producing a coarse sonar map. The sonar map would be used to plan where to deploy the ROV, which would then use its sonar to navigate to the suspected area and then the camera to see the object. However, the AUV had a malfunction and could not be repaired in the field. Instead, the ROV was used at areas of interest indicated by the fishermen's depth sonars.

Despite the limitations in staging areas and the inability to use multiple robots concurrently, the ROVs were effective and fast. In the April deployment, the robots searched approximately 2,400 square meters in 9.25 hours of time in the water, far faster than manual divers could work. The Minamisanriku "New Port" area, vital to reestablishing the local fishing economy, was reopened based on the search for debris in turbid waters that might damage a fishing boat or foul propellers. The ROVs were able to confirm to families that their missing family members were not underwater in several places that the Japanese Coast Guard could not dive in. In the October deployment, the ROV was able to cover at least 80,000 square meters in 6.2 hours averaging 217 square meters per minute—a significant increase in coverage due to the use of a boat. On the first run in October, the team was taken to a fishing area that had been investigated by manual divers and declared free of debris and asked to show how the ROV worked. The ROV found a sunken commercial fishing boat, showing the difficulty that manual divers have in reliably searching in turbid waters.

5.5   Selection Heuristics

Unlike UGVs where the primary selection attribute is size, there is no single criterion for selecting a UMV for a mission. The case studies described earlier illustrate the multiple influences on selecting the appropriate UMV for an event. Below is a summary of heuristics used by CRASAR:

- If data are needed in real time, either to direct the robot (e.g., "get a view from this position") or to intervene (e.g., recover the victim from a car or turn a valve), then a USV or ROV is a better choice.
- If the access to or at the site is limited, an ROV will most likely be the easiest to launch and recover from limited space and is at lower risk of being lost than an AUV or USV.
- If the mission requires operating near structures, USVs and ROVs can be visually guided, and USVs can use GPS.
- If the mission requires coverage of a large area, AUVs may be a good choice depending on the amount of debris.

• If both video and sonar are needed, then an ROV or USV is a good choice.
• If the mission is bridge inspection, a USV can inspect both the structure and the upstream debris field.

5.6   Surprises, Gaps, and Open Research Questions

Possibly the biggest surprise about unmanned marine vehicles is that they have a significant role to play in disaster response and recovery, followed by the ways they are used. One major set of missions is for littoral inspection of built structures such as bridges and ports. A surprising aspect of this inspection mission is that it includes mapping the upstream debris field that may damage a bridge at a later date or prevent entry into a port. Another surprising aspect of inspection is that inspecting the portions above the waterline is of little interest, so UAV—UMV combinations for bridge inspection are not particularly valuable. A technical surprise that is a gap between what is available and what is needed is that GPS and line-of-sight wireless coverage is not perfect when working near these structures; despite the perception that a UMV is working in open space, bridges and ports create GPS shadows and effects similar to the shadows and effects experienced by UAVs operating in urban canyons. Another surprising gap is the difficulty of accurate control: The effects of currents, wind, and tides are often underestimated. Indeed, roboticists often naïvely assume that the control dynamics of USVs are similar to those of UGVs—neglecting that a USV keeps moving due to momentum and, even if no motors are active, may drift off course.

One set of gaps that must be addressed by research and development stems from the two contradictory design requirements: UMVs are produced commercially primarily for open-water applications, while UMVs for search and rescue typically operate in littoral regions. UMVs are a mature commercial sector, where robots are routinely used for oil rig inspection (ROVs) and marine biology (UUVs). These ROVs are used for routine tasks by highly trained operators, but disasters are the antithesis of "routine," and responders will not be able to maintain the high level of training needed to avoid human error. UUVs work in open water where there are no obstacles, and changes in depth due to tides typically do not matter for a disaster mission, but littoral regions are the critical areas that require UMVs.

There are two informatics gaps in UMVs. One is that the responders have to have confidence in coverage—that the UMV did in fact cover the area and nothing was missed. This confidence can be produced qualitatively through perception (e.g., a three-dimensional rendering of the area that explicitly shows what has been searched). However, more work is needed in statistical methods to ensure sufficient overlap of observations for the positional accuracy or to determine if any areas or viewpoints were missed. A second informatics gap is the lack of computer vision for interpretation of imagery. Only one robot, the SARbot, had any type of computer assistance—in that case, the LYYN image enhancement software, which improves video image quality in turbid waters. Other applications of computer vision such as three-dimensional reconstruction, highlighting edges, identifying regions based on differences in texture, and so forth, are missing.

The relatively high number of human errors in UMV deployments and their consequences (e.g., damaging remediation devices, risking the UMV, etc.) indicate that human–robot interaction remains a large gap. One topic is how to generate sufficient situation awareness, especially for ROVs using tethers in cluttered areas. In general, user interfaces do not facilitate an operator or structural expert being able to know intuitively where he or she is looking or where the robot is (or has been).

Effective deployment of multiple robots requires more research and development as well. Deploying multiple robots with active sonar requires significant coordination due to the possibility of interference between the sonars. Likewise, ideal divisions of areas may be affected by the lack of staging areas. Using UMVs in conjunction with UAVs and UGVs has begun to be explored, but much work is needed to transfer those results to the extreme physical and network connectivity conditions of a disaster.

5.7 Summary

UMVs provide remote presence underwater, either from true UUVs—tethered ROVs and untethered, free-swimming AUVs—or from USVs. Each type of UMV presents unique challenges for control, perception, and human–robot interaction. All but one of the UMVs used in deployments have been commercially available vehicles rather than laboratory prototypes, probably due to the complexities of manufacturing a platform that can work reliably underwater and in corrosive saltwater.

Returning to the three design constraints on rescue robots, UMVs *experience extreme environments and operating conditions, must function in GPS- and wireless-denied areas,* and *pose extreme human–robot interaction challenges.* UMVs work primarily in the extreme terrains of shallow littoral waters but can also be used in deep water. The size of the UMVs used to date has been constrained by transportation concerns (e.g., how to get it to the site) and ability to work in shallow water. However, the size of any structures being investigated (e.g., an intake/outflow pipe, spacing of pilings, or density of debris) may impact future deployments. UMVs for the rescue and initial recovery phases must be operable without a boat and require a small staging space. UMVs that work underwater either have no access to GPS and must rely on inertial sensors or use a tether and transitive methods for localization. UMVs typically rely on acoustic imaging rather than video as their primary sensor in order to function in turbid waters, requiring greater expertise for interpretation. The use of sonars and video, which have different formats and ranges, adds to the considerable human–robot interaction challenges of using UMVs. Human operators often lose track of ROV tethers, which tangle. Whereas human error accounted for 50% of UMV mission failures, 33% stemmed from inability to work in the current or turbidity and 17% from damage during shipping.

The seven reported deployments show that UMVs can rapidly survey and inspect littoral built structures and the surrounding debris field in far less time than manual divers (in a few hours with setup times between 10 and 45 minutes) and with greater accuracy. They offer value for *structural inspection; estimation of debris volume and type; victim recovery; forensics; environmental remediation; search, reconnaissance, and mapping;* and *direct intervention.* UMVs do not replace manual divers for debris removal missions but can locate debris for the divers and reduce their time in the water.

As with UGVs and UAVs, the biggest gaps in UMVs are sensing and human–robot interaction, though the need for manipulation is increasing. UMVs follow the trend of requiring the operator mentally to construct and maintain situation awareness from separate raw sonar, video, and positional data feeds. AUVs do provide a tiled sonar map, but interpretation of the sonar imagery requires significant expertise, which may be beyond the capabilities of first responders. The only real computer support witnessed to date has been the use of video image enhancement for turbid waters.

In the future, UMVs may make the biggest contribution to prevention, preparation, response, and recovery from disasters. UMVs provide sensing and action in environments that otherwise require highly specialized workers in restrictive protective gear. Littoral regions contain critical infrastructure that may be the target of terrorists or if damaged by an event could retard the arrival of responders and relief supplies. New types of low-cost UMVs are coming on the market almost daily, and advances in autonomous capabilities and in underwater communications will accelerate the real-time utility of these devices.

6   Conducting Fieldwork

The previous chapters introduced rescue robotics, provided discussion of where the robots have been used and for what types of missions, and included descriptions of ground, aerial, and marine robots. This chapter offers suggestions on how to conduct research in the field and on deployment to a disaster. The chapter can be used by disaster professionals for evaluating disaster robots and as a guide for designing field exercises and collecting meaningful data at deployments. It can also be used by industry professionals for creating effective testing and for determining what type of data logging should be incorporated into their robots.

By the end of this chapter, the reader should be able to:

• Understand the difference between fieldwork and laboratory experimentation in terms of **authenticity**, **quantitative measurability**, and **repeatability** and understand the importance of **holistic evaluation** and the dangers of relying on tabletop exercises.
• Be familiar with the differences between **controlled experimentation**, **participating in exercises**, **conducting concept experimentation in exercises**, and serving as a **participant-observer** in a **deployment** in terms of *venue*, *research driver*, and *designer*.
• Know the five basic facets of a field event to consider in planning: **fidelity, involvement of response professionals, logistics and tempo, risks, and safety**.
• Know the roles of team members, especially the **robot expert operator, mission specialist**, and **field safety officer**.
• Know the six categories of data to try to collect (**log of activity, context, robot's-eye view, robot state, external view of the robot, human–robot interaction view**) and the essential elements of a **data collection protocol**.
• Understand how to participate in responses.

The chapter begins by discussing the motivation for conducting research and development in field conditions (where the physicality and the sociotechnical complexity of disasters are preserved) and the trade-offs between authenticity, measurability, and repeatability between fieldwork and controlled experimentation. Next, it describes the four types of field-work: controlled experimentation, participating in exercises, conducting concept experimentation in staged worlds, and being a participant-observer during a deployment in the natural world. The chapter then details the five basic areas that a team leader needs to consider in planning a field event and the 10 distinct roles that must be played by a team that may have less than 10 members. There are four questions that must be answered in order to determine if robots of any kind are a good fit for an event. The schedule in the field is also discussed, and a data collection protocol is outlined, though it is difficult to gather all the data that will be helpful for truly understanding the performance of the system. The chapter concludes with observations on the etiquette and psychological concerns of participating in a disaster exercise or mission.

6.1 Motivation

Fieldwork is an essential element of research and development of disaster robots. As seen in earlier chapters, disasters pose extreme environments that are hard to duplicate in a traditional indoor laboratory setting and involve large numbers of professionals interacting with the robot(s) or the data from the robot(s). Thus, it is important to know the types of fieldwork and trade-offs between the predominately qualitative output of fieldwork and the quantitative output from laboratory experimentation. Fieldwork is also necessary for holistic evaluation of a robot in a larger sociotechnical system, where performance includes how well it fits the constraints of the existing workflow.

6.1.1 Differences between Fieldwork and Laboratory Experimentation

Figure 6.1, which is adapted from Woods and Hollnagel (2006), illustrates the difference between fieldwork and laboratory experimentation by plac-ing the activities in a larger context of what are the sources of data for under-standing systems. The mechanisms for understanding systems are divided into three categories. At the bottom level is the *corpus of cases about the natu-ral world*, which is obtained by direct observation in the field. For example, Isaac Newton could observe apples falling. As in Newton's case, these obser-vations form the basis for the development of models and hypotheses that would be the subject of focused experimentation, either in a traditional *controlled laboratory setting* or a *staged world* venue that simulates a broader set of influences. A **staged world** is a physical simulation that replicates a set of key features of the natural world with high fidelity; it is often a field exercise or experimentation conducted in the field, but the actual activities are constrained so as to be repeatable. Perhaps one of the most memorable staged world observations was Galileo dropping two balls of different mass from the top of the Leaning Tower of Pisa, which, in turn, was pivotal in getting Astronaut David Scott to the moon where he dropped a feather and a hammer at the same time, showing both of them falling at the same rate in the natural world. Staged world observations simulate the natural world, usually by attempting to replicate a mission and operational environment for a robot. A controlled laboratory setting is not the field, but rather captures one or two aspects of the field in a highly replicable setting such as a computer simulation or a physical mock-up.

Each of the three categories of acquiring understanding has a different degree of *authenticity, quantitative measurability,* and *repeatability.* The natural world is by definition the most authentic representation of the natural world and no matter how well an idea or product tests in the lab, it is how it behaves or is used in the real world that matters. But the natural world is messy: The overall system, in this case a disaster, is nonlinear; has large

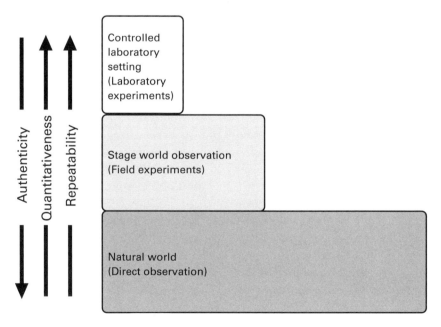

Figure 6.1
Mechanisms for understanding systems. (Adapted from Woods and Hollnagel, 2006.)

interdependencies between the environment, the responders, the victims, and so forth; and has large temporal and geographical scales. The natural world is generally not repeatable, both in physical conditions and the sociotechnical aspects. Good performance of a robot in one hurricane does not necessarily mean good performance in another hurricane because so many factors can be different. The lack of repeatability interferes with assessment and with the ability to quantify the assessment. On the other end of the spectrum, lab experiments sacrifice authenticity in order to achieve repeatability and quantifiable results. Staged world observations appear to occupy the middle ground, but the quality of the research depends on how well the designer balances the fidelity of the simulated experience with the need to control variables and the ability to observe sociotechnical interactions.

Figure 6.1 also reflects an implicit ordering of gaining understanding about a system. The bulk of the understanding should be from the natural world, which provides the motivation and data for creating models and hypotheses that can be tested either in a controlled laboratory setting or in a staged world. The focus on hypothesis-directed research in laboratories in medicine, physics, and many engineering disciplines is the result of hundreds of years of observation about the natural world.

Unlike disaster robotics, many of the sciences have a large corpus of prior observations, and thus ideation may not require new fundamental observations of the natural world. Instead, there is an implicit foundation for ideation and for positing a hypothesis to prove it. Researchers in fields such as aerospace engineering can begin with laboratory experiments immediately, scale up the work to staged world observations, and then confirm the results in the natural world with new observations. This hypothesis-driven research approach is referred to as **scientific reductionism** because controlled experimentation requires reducing the scope of the hypothesis and the world so that the central effect can be observed and the experiment repeated to achieve statistical inference. As seen in chapter 2, disaster robotics has a very small corpus of natural world observations, suggesting that laboratory experimentation or even staged world experiments will have limited utility as there is little information to create meaningful models or to know what aspect of the natural world to duplicate.

6.1.2 Holistic Evaluation

Computer science, robotics, and especially rescue robotics are new fields driven by the creation of new capabilities, which require **holistic evaluation** of the technology in the larger sociotechnical system. Holistic evaluation is typically expressed as "Did it work or didn't it?" As disaster robots are part of a complex sociotechnical system, this question becomes harder to answer; for example, *if a new sensor returns high-resolution data but the operator makes significant errors because of the way the data are displayed, is the new sensor really a success?* Holistic evaluation is generally performed with staged world observations or with observations in the natural world.

Holistic evaluation means that disaster robots have to be tested in a staged or natural world. Access to an actual disaster is rare and, as will be discussed later, disasters are not the place for experimenting with the untried. This means that most holistic evaluation will have to take place in a staged world. Staged worlds are difficult to create because there is a relatively small corpus of cases where robots have actually been used to serve as models. As a result, it is not clear how to answer questions such as: *What is the right scale and terrain needed for sufficient fidelity? How many and what type of responders are needed, just the operator or also representatives of the command hierarchy? Is it important to test for multiple days?* In general, computer-based simulations do not provide the sociotechnical system needed for holistic evaluation.

Tabletop exercises (TTXs) are often an initial step in holistic evaluation as they bring together experts to walk through scenarios, providing the socio-organizational aspect without the overhead of a fielding of the technology. For example, I observed a TTX hosted by AUVSI that brought together UAV manufacturers and wildland firefighting agencies. A major conclusion was that the one feature that would entice responders to adopt existing small UAVs was the delivery of data (Murphy, 2010b). The majority of manufacturers represented were primarily recording low-resolution video to a USB drive and still imagery to an SD card on the platform. From a technical perspective, this was acceptable, but from a socio-organizational perspective, it was not. The recording scheme meant the responders had to wait for the asset to return before obtaining high-resolution photographs. They also had to bring a separate laptop to review the video USB drive. The responders wanted to see the images and be able to replay the video while the UAV was still in the air to be sure that they captured the desired data, thus saving time, reducing the number of flights, and reducing the risk that smoke, weather, or other conditions might prevent the next flight.

However, tabletop exercises are generally only a first step, and findings can be heavily influenced by the choice of attendees, the way the problem is framed, and the assumptions made. We conducted concept experimentation with hazmat instructors from across the country in a series of chemical, biological, radiological, nuclear, and explosive (CBRNE) staged world events (Murphy et al., 2013b). Concept experimentation obtained a totally different concept of operations and human–robot interaction than a CBRNE TTX and survey conducted by Humphrey and Adams (Humphrey, 2009; Humphrey and Adams, 2009). Their TTX concentrated primarily on strategic decision makers and assumed that the region had dedicated robotic CBRNE assets and communications; not surprisingly, the big problems were how the city administrators could visualize the event represented as icons on a map. Our work did not impose assumptions on the scenario and discovered that it was unlikely that fire rescue would have dedicated CBRNE assets and that the major decision making using robot data would be at a tactical level. These discoveries suggested a very different set of issues in a CBRNE event, as the robot would be part of an ad hoc team consisting of probably a law enforcement pilot and a fire rescue mission specialist that may have never worked together. Furthermore, there was an average latency of nearly 30 minutes in getting the data from a UAV or UGV to the tactical decision maker.

6.2   Four Types of Fieldwork

For the purposes of this book, fieldwork is formally defined as *research conducted either in a staged world with significant physical fidelity* (e.g., in rubble at a fire training academy, a building being demolished, etc.) *or in the natural world* (e.g., a disaster site). There are four types of fieldwork for obtaining data about disaster robotics: *controlled experimentation, participation in an exercise, concept experimentation,* and *participant-observer*. These four types are differentiated by the venue the work occurs in, the driver or objective of the work, the designer of the experience, and the expected outcomes. Table 6.1 summarizes the four types of fieldwork.

6.2.1   Controlled Experimentation
Controlled experimentation is highly focused, either on testing a hypothesis or capturing a performance metric(s) such as cognitive workload. Although similar to lab experimentation, it takes place in a staged world. The fidelity of the staging depends on the particulars of the experiment. Like lab experimentation, the complete activity is designed and controlled by the researcher to maximize meeting of the research objectives (figure 6.2).

An example of controlled experimentation in a staged world is the study conducted by one of my former students, Dr. Josh Peschel, in human–robot interaction. Based on the findings of my group's human–robot interaction research regarding how teams of responders and operators collaborated using the shared visual perception protocol with UGVs (Burke and Murphy,

Table 6.1
Four types of fieldwork in terms of driver or objective, venue, designer, and outcomes

Type of Fieldwork	Driver	Venue	Designer	Outcomes
Controlled experimentation	Hypothesis or performance	Staged world	Roboticist	Statistically valid inferences
Exercise	Successful demonstration	Staged world	Stakeholders and roboticists	Familiarity, favorable opinion, accelerated adoption, some feedback
Concept experimentation	Mission suitability	Staged world or natural world	Stakeholder	Gaps analysis, possible new uses, concepts of operations
Participant-observer	Authenticity	Natural world	Stakeholder	Case studies of actual uses and concepts of operations, failures and bottlenecks

2007) and UAVs (Pratt, 2009), Josh hypothesized for his doctoral thesis that responders working with a UAV pilot to perform reconnaissance would do better with their own dedicated display that also maintained a visual common ground with the pilot. To test this, Peschel conducted research at Disaster City® using the highly realistic chemical train derailment prop (literally a real chemical train, derailed). His human subjects were all trained hazardous materials response professionals, providing a high degree of personnel fidelity in addition to the perceptual fidelity. But in order to test his hypothesis and make statistically valid inferences, he had to limit the tasks the responders were allowed to do with the UAV and fix the order in which they collected data in order to create a repeatable test.

6.2.2 Participation in an Exercise

Participation in an exercise occurs in a staged world. Most exercises concentrate on reinforcing good practices, so these often focus on everyone successfully performing their roles. Exercises are success driven, where the event is designed to maximize demonstration of the favorable contributions of the technology. The physical environment is often made safer than what would be expected in the natural world so as to reduce health risks. Response exercises are *scripted by participating technologists and responders in order to demonstrate the benefits of a technology*. Responders have to use the technology as planned, and monitors may intervene to give instruction

Figure 6.2
An example of fieldwork for controlled experimentation: A responder uses a dedicated display under controlled conditions to view UAV imagery of a simulated chemical train derailment.

or the exercise may be temporarily suspended while a computer reboots, the wireless network has a problem, and so forth. As a result of scripting, participation in exercises is rarely deeply informative about performance or acceptance. Fortunately, the robots will usually trigger the response professionals to ask questions and make novel suggestions that can be the subject of further research (figure 6.3).

An example of an exercise was CRASAR's participation in the 2007 Community Integrated Disaster Response Exercise (CIDRE) at Wichita, Kansas. This was a full-scale exercise that ran for 48 hours with more than 600 participants drawn from the multiple agencies that would be involved in a terrorist event. The exercise was unusual in both the number of people involved and the physical realism—the event took advantage of buildings being demolished in different parts of town. CRASAR members provided UGVs for one of the technical search components that more than 100 participants had to complete in order to earn a certificate, requiring redundant robots and tools and expertise for rapid repairs. However, the event did

Figure 6.3
An example of fieldwork at an exercise: one of the venues for the 2007 CIDRE exercise held at Wichita, Kansas.

have an opportunity for a UAV and a UGV to be used in a more free-form style by a SWAT team tracking a suspect as he entered a building.

6.2.3 Concept Experimentation
Concept experimentation usually occurs in a staged world but opportunistically in the natural world where the robot is performing a mission. The experimentation is focused on generating concepts of how a new technology or protocol can be used, leading to what is called a **mission capability package** (Alberts and Hayes, 2002). Even though a developer has some expectation or intent of how a new technology would be used, new technologies always present surprises when given to the end users. A hallmark of concept experimentation is that the robot is allowed to fail in the staged scenario or activity; indeed, the activity is expected to be complex and hard enough to cause hardware or software failures and thus uncover performance gaps and human–robot interaction problems. Concept experimentation also identifies new uses or missions for the robot, making it particularly appropriate for rescue robotics because, as described in chapter 1, the field is **formative**.

While the majority of concept experimentation occurs at a fire rescue training facility or military test range, an example of concept experimentation in the natural world was the use of a USV to inspect the underwater portion of the Rollover Pass Bridge that was damaged by Hurricane Ike in 2008 (Murphy et al., 2011b). With permission and engagement from the Texas Department of Transportation, CRASAR deployed the Sea-RAI unmanned surface vehicle and two other underwater vehicles (figure 6.4). The intent was for a team of experts in bridge collapses to evaluate the robots and whether they were practical (or could be in the future) for bridge inspection. The civil engineering experts spent 3 days determining if the bridge pilings had signs of scour or if there was debris upstream that could endanger the bridge. This was a situation where there were missions, but there was time and access for re-running missions to explore how these robots might be used and the value of the data. An outcome was the determination by the civil engineers that the robots were useful, and the roboticists collected performance data and documented the loss of GPS signal near the bridge. The natural world presented logistical challenges, as there was limited access to boat ramps, roads, electricity, housing, and gas.

Figure 6.4
An example of fieldwork for concept experimentation: a USV at the Rollover Pass Bridge, which was damaged by Hurricane Ike.

6.2.4 Participant-Observer Research

Participant-observer research is conducted while the robot is actually deployed to a disaster, and is a form of **ethnography**. The deployment is focused on contributing to the larger response organization, which may mean deciding not to deploy a robot. Data may be collected only as circumstances permit. Participant-observer research is authentic but often consists solely of ethnographic observations, the quality of which depends on the skill, and amount of sleep, of the participant-observer.

An example of participant-observer research is CRASAR's insertion of robots at the September 11 World Trade Center collapse (figure 6.5). UGVs were on the scene and being deployed within 6 hours of the collapse and were used for nearly a month, yet this showed how hard it is to collect data. New York City prohibited taking photographs out of deference to the victims, so there are few context shots of where the UGVs were being used. The team members had less than 3 hours of sleep over the first 52 hours, video data from the first day was accidently taped over, and notes in written journals veered into incomprehensibility. But while the data gathered were incomplete on some levels, it was authentic, and the analysis in Casper and Murphy (2003), led by my student Jennifer Casper, is considered a seminal paper for human–robot interaction.

6.3 Planning for Fieldwork

Regardless of the type of fieldwork and goals, there are five considerations that will influence the quality of the experience and how smoothly it will unfold: the *fidelity* of the event; the degree of *involvement of response professionals*; planning of the *logistics to support the tempo* of the event; anticipating and *minimizing the risks* to the robots, researchers, mission, and the future adoption of disaster robotics; and ensuring of *safety* during operations.

6.3.1 Fidelity

Having access to a field event with sufficient fidelity is the single most critical factor in obtaining research results. The goal for controlled experimentation, participation in an exercise, and concept experimentation is to determine the level of authenticity necessary to obtain meaningful results. In general, a roboticist will be concerned with the *physical fidelity of the site* (Is the terrain similar to the terrain a robot would be used in?), the *mission fidelity* (Does the use of the robot mirror the expected tasks or mission and operational procedures that would be used in the field?), and

Figure 6.5
An example of participant-observer fieldwork: robot operator Arnie Mangolds inserts
a UGV into a void at the World Trade Center collapse.

socio-organizational fidelity (Is the event using the type of personnel and
roles that would be used in a disaster?).

A common mistake frequently made by roboticists is to go to a fire training academy and run their robots *over* a concrete rubble pile as a test of
mobility instead of *in* the interior. A variant of this mistake is for the robot
to enter the interior of the rubble using horizontal passageways; these passageways were installed to make it easier and safer for humans who are
pretending to be victims, not to be representative of voids. As noted in
chapter 1, rescue robots are generally used in interior spaces too small for
a human to enter and that have significant vertical traverses. Furthermore,
roboticists might become overly optimistic about performance of a wireless network if the metal concrete-reinforcement bars ("rebar") have been
removed from the concrete rubble. This is an example where the physical
fidelity (the large horizontal voids) and the mission fidelity (running over
the rubble, not in it) are poor.

*There is often confusion between a test for a capability that relies on a high-
fidelity simulation of the critical aspects of the site and the mission fidelity.* For
example, each NIST Standard Test Method focuses on testing a particular capability in as realistic but repeatable conditions as possible, such as

navigating a three-dimensional maze. But the test methods do not capture the overall mission fidelity, which includes how the robot gets in and out of the void, sensing, human factors, and so forth.

Another common mistake is to ignore mission constraints, such as stand-off distances. The robot operators may be operating 0.5 kilometer away from the robot.

6.3.2 Involvement of Response Professionals

Response professionals will most likely be involved in fieldwork because you will be using their facilities, working under their direction at a disaster site, working alongside them as participants, or will be collecting data about them for human–robot interaction.

If you are using their facilities or working under their direction, they will be in charge of supervision and safety. You should ask and make sure that you understand their rules of operation and safety practices if they forget to tell you.

If the response professionals are in charge of the fieldwork, such as running an exercise or the response, there should be a single point of contact— the person you report to. Making sure that that person understands the objectives, the equipment, and what you are trying to do is imperative, so that you both can converge on a workable plan. Sometimes the answer will be "No, you can't do that." But that is better to know before the field event than in the middle of it.

If your research involves responders directing or using the robot, you will have to create a training plan and determine when the responders would do that. The training plan may be short. For example, at a recent controlled experiment at Disaster City® with unmanned aerial vehicles run by Josh Peschel for his doctoral work, each responder began the test session with a 10-minute practice flight at the staging area before he or she began the actual long-distance flight. In contrast, the training plan may be more involved. For Jenny Burke's research, responders took a 1.5-hour course consisting of classroom introduction and field practice the day before the actual event. The course was sufficiently formal that responders received continuing education credits.

6.3.3 Logistics and Tempo

Logistics and tempo must be considered when planning fieldwork. Logistics primarily is concerned with making sure everything needed for fieldwork arrives and is in the right place. The tempo of the fieldwork influences the amount of resources (e.g., batteries, recorded media, etc.) needed to meet the pace.

Roboticists often forget that logistics isn't restricted to transporting the materials to the site. Once at the site, the gear has to be split up and transported (usually hand carried 100 meters or more) to the actual forward operations base in the hot zone. Wheeled cases that worked well in laboratories and airports don't roll as well over dirt and rubble. The cases are often difficult to carry up or down hills of rubble. Changing to a backpack is an option, but care must be taken that the gear doesn't interfere with the person's center of gravity and put the individual in danger while he or she is climbing through the rubble.

Another aspect of logistics is organization of the gear. It can be worth the effort to have everything needed for a robot in one or two cases, separate from recording gear or other equipment. We've had several events where "efficient" packing led to the event having to wait. In one event, one of the cases was delayed by the airline, that case naturally had a key part for the robot. A similar situation is when in the rush to send a robot to area A and another to area B, one group doesn't gather everything it needs because it was all packed together.

Logistical considerations for staged world situations include infrastructure, especially power, which is influenced by the tempo. How many batteries are needed? Is a generator required?

In a natural world situation, more basic infrastructure—food, water, sanitation, power, gas, places to sleep, and sometimes security—are usually not available. Response agencies aggressively discourage unsolicited volunteers and "disaster tourism" because they have typically brought only enough for the people they are directly responsible for. A robot team may add capability, but it also presents a logistical challenge for the response agency, which has to be sure there are enough resources to add the team to its roster. The less you impose on the responders, the more likely you will be involved in the response.

The tempo of a field event, exercise, or disaster is variable, and robots and roboticists must be aware of the expectations. The tempo affects robot usage. At the World Trade Center, responders expected the robots to provide video imagery as fast as their search cameras; they would literally walk away and not return due to the slow computer boot times. We quickly learned that the systems had to remain on hot standby. At Hurricane Katrina, we discovered that the initial expectation that UAVs would need to remain in the air for more than 20 minutes was incorrect. The necessary views were generally found within 8 minutes; the problem was not having sufficient battery life but rather wasting time discharging the battery so that it could be recharged to peak efficiency.

The tempo also affects people. We have participated in several exercises that ran for 24 hours continuously. This meant that team members had to be scheduled for rest periods. At the World Trade Center, CRASAR members had less than 3 hours of continuous sleep in the first 52 hours of the disaster—a number that every US&R specialist I've ever spoken to says is typical.

6.3.4 Risks

Every activity has its risks, and fieldwork is no exception. There are three categories to consider: *risk of injury to the researchers or loss of equipment*, *risk to the mission*, and *risk to the future adoption of robotics*. To combat these risks, roboticists must:

• Manage the responders' expectations of what robots can do, avoiding any trace of hype or salesmanship. This includes realistically assessing the potential for success or failure of a particular robot. This is fairly straightforward to do as it entails the team simply being aware of the limitations of their devices and being up front, honest, and at all times conservative in discussing the robots.
• Bring a team that is familiar with safety rules, can work well and with each other under pressure, and evinces the greatest respect for the responders as professionals. Not every roboticist is good in the field, just like every doctor is not necessarily good in the emergency room. This requires that the team practice with the robots, each other, and responders.

The first category of risks is that the use of robots in high physical fidelity field conditions with responders poses risks to the equipment and personnel; people or equipment can get hurt. Personal safety is of course the most important issue, and it is the responsibility of the research lead to make sure that all team members are aware of and adhere to safety standards, have appropriate safety gear, and so forth. Safety generally requires training and practice.

While personnel safety is critical, it is also important consciously to decide in advance how much risk to the equipment is tolerable and how to mitigate the risk. The members of CRASAR's Roboticists Without Border's program understand that during a disaster deployment, their robot might be lost or damaged and might not be reimbursed. A Foster-Miller Solem robot was lost and not recovered during the World Trade Center response. At the Midas Gold Mine collapse, the Inuktun Xtreme robot from the Naval Sea Systems Command and Space and Naval Warfare Systems Command (NAVSEA SPAWAR) at San Diego's small mobile robot pool was damaged by falling rock. In assisting with the forensic structural inspection of a portion of a collapsed parking garage that was still standing, we were unable to enter the structure if necessary to retrieve the robot. As a result, we chose to send in a tethered robot because it would not have wireless communications issues and so if got stuck, we could try to pull it out and would not have to worry about battery power (i.e., we could take a break and try it later or the next day, a strategy that worked at the Crandall Canyon Mine collapse). However, equipment risk is minimized during controlled or concept experimentation; for example, we will not fly in certain difficult locations or near structures without spotters because we can't afford to lose the robot for the remainder of the event (or for our research program in general).

The second category of risks is that the robot itself can pose a risk to the entire mission if it behaves unexpectedly or fails—and it is the researcher's responsibility to make realistic projections as to what the robot will actually be able to do. As seen at the Pike River Mine collapse in New Zealand, the failure of one of the rescue robots blocked entry into the mine for several hours and made the situation worse. While preparing for the Crandall Canyon Mine collapse deployment, I identified and presented to the Mine Safety and Health Administration seven distinct ways in which the robot could fail and thus end the deployment. In the field, we encountered every single projected failure mode; we were able to recover from six of them but not for the seventh, a cave-in, which fortunately happened as the robot was exiting the borehole from its last run.

In general, robots perform worse in the field than in the lab, which is reasonable to expect given the extreme physical challenges of a disaster environment. My group's documentation of how robots fail in the field (Carlson and Murphy, 2005), including autonomous software issues, led Dr. David Woods to create in 2006 (Woods and Hollnagel, 2006) *"(Robin) Murphy's Law: Any deployment of robotic systems will fall short of the target level of autonomy, creating or exacerbating a shortfall in mechanisms for coordination with human stakeholders."* As Woods and I later discuss in Murphy and Woods (2009), **operational morality** dictates that roboticists be aware and up front about their robots and not use "who knew it could fail this way so we didn't have a backup plan" as an excuse. Robots will fail or encounter unpredictable situations. Robotics designers can design systems and use protocols that enhance our ability to deal with surprises.

The third category of risk is that the performance of the robots and the behavior of the research will negatively affect the perception of rescue robotics and researchers by the response community. Roboticists without significant field experience typically overstate the robot's capabilities and do not recognize that responders are themselves highly trained professionals whose professional commitment is to give their lives to save ours.

A consistent problem has been "vendor-itis," my term for when robot manufacturers and researchers use marketing tactics—especially overstating the capabilities and neglecting to mention problems—when interacting with responders. (A related problem has been approaching the responders during a disaster and trying to get them to pay for using the robot—this will be discussed later in this chapter.) Of course, there are cases where roboticists have overstated capabilities in honest ignorance, but the result is the same: When robots underperform or roboticists act like salesmen, the responders form a negative opinion of all robots and roboticists.

A second type of behavioral problem has been a failure to treat responders as professionals or to adhere to their directives. One heart-stopping example was at the World Trade Center disaster. We had just gotten the FEMA urban search and rescue teams interested in the robots—no easy feat given that new technology is rarely adopted during a disaster—and were demonstrating them to a small group at the base of operations. A member of the CRASAR team proudly told team leaders that these robots would one day replace them, leaving me to try to smooth over the deservedly angry reaction. Another frustrating instance occurred at a CRASAR exercise with Indiana Task Force 1. Indiana Task Force 1 had arranged for us to have access to a building that was being demolished, and we brought in about 25 roboticists to spend 2½ days of concept experimentation. The responders reviewed safety practices with us, emphasizing that once inside the surrounding fence, we had to wear hard hats. Despite the safety briefs and constant reminding, one roboticist stubbornly kept entering the area without his hard hat, which gave the responders a poor impression of how roboticists could be counted on to behave in the field.

6.3.5 Safety

Planning for fieldwork must include safety. The team should always have a safety officer and a safety plan. The safety officer for a squad holds only the safety officer role so that he or she does not get distracted. The safety plan should consider how the team will egress from the site. For example, at the World Trade Center collapse, the robotics gear could not be quickly disconnected from the operator control station carried in a backpack by the operator; thus, the plan was for the operator to remove the backpack and leave all the gear behind if the evacuation signal was given (e.g., imminent collapse). At the Tohoku tsunami response, the safety officer made sure that there was a car parked nearby and the driver was familiar with the route to flee another tsunami.

In most fieldwork, such extreme considerations are not necessary, and simple rules are sufficient. Each team member should dress properly, with long pants, long sleeves, and a hat or hardhat, and closed-toe shoes unless steel-toed boots are required. Safety glasses, hearing protection, and gloves may also be appropriate for different venues. Team members should adhere to the *buddy system* of working in groups of at least two and being responsible for each other, and each squad in the field should have sufficient water, a first aid kit, sunscreen, bug spray, and so forth. Members should be on the alert for heat stroke or hypothermia and also aware of the signs of cognitive fatigue.

The importance of sanitation and decontamination is perhaps the most underestimated aspect of safety. There is the need for personal sanitation, especially to wash hands before eating given portable (or nonexistent) restrooms. Proper food handling and storage for long field events are important as well. The robots and team member clothing, especially boots, need to be cleaned off and decontaminated. In an exercise, this keeps everything clean and keeps dirt out of sensitive robot parts. But in a real deployment where a building has collapsed, the site will be covered with sewage or even body fluids. It is handy to pack with each robot a brush, a can of compressed air (to blast the dirt out of crevices in the robot), a plastic garbage bag to put a contaminated robot in, and a towel (*The Hitchhiker's Guide to the Galaxy* was right).

6.4   Organization of Robots and Team Members

A major element of successful fieldwork is selecting and organizing the robots and team members. While every field event is different, this section identifies 10 key roles that have to be filled by team members regardless of type of field event, discusses the need to maximize diversity and redundancy in the choice of robots and gear, and describes typical scheduling.

Our past experiences can be distilled into a set of guiding principles:

• Bring the smallest group of people and robots possible to reduce the demand for food, water, sanitation, and transportation at the disaster.
• Never bring robots that have not been tried extensively before in similar conditions.
• Never bring people that have not worked in the field or with each other before.
• Specify and enforce a clear chain of command and responsibilities.
• Keep setup simple: Assume that if a connection can be made incorrectly, it will be, and thus keyed connectors, color-coded connectors, and labels should be used.

• Practice, practice, practice.
• Mentally visualize the entire process, from walking out the door, to conducting each mission, to returning.

6.4.1 Roles

There are at least 10 distinct roles held by members of the team performing the fieldwork: **Team leader, technical lead, field lead, team public information officer, robot expert operator, mission specialist, team safety officer, field safety officer, data manager,** and **scientific observer**. As the general principle is to minimize the number of people in the field, individuals typically hold multiple roles; for example, the team leader may double as the team public information officer. Individuals also may switch or take turns at roles at events where a squad of members are sent with robots to the field and other members stay behind or form another squad; for example, the technical lead is almost always a robot expert operator and field leader in a squad.

This section describes each role in terms of general responsibilities before, during, and after a field event, but there are three important caveats to consider. First, *some of the roles may be filled by responders*; the most common of these appear with an asterisk (*) in the list that will follow. In most CRASAR deployments, the hosting agency provided at least one response professional to serve as the field safety officer or as the mission specialist. This means that the team should have some plan for explaining the processes or conducting **hasty training** in the field. Second, the large number of roles indicates how complex even participating in an exercise is, and *the more shared roles an individual holds, the more likely he or she is to make mistakes in operations, data collection, and safety* if there isn't practice. Third, *not everyone is well suited for fieldwork or for working with responders*. The previous sections have illustrated that individuals may not be safety compliant or respect responders, both of which are essential. Another measure of suitability is whether an individual understands that all tasks are essential and not everyone will be "center stage"; for example, data management back at the base of operations is as important as operating the robot in the field.

• **Team leader** The team leader is the overall leader of the entire group. Before the field activity, the team leader is the point of contact with the agency, embassies, and so forth. The team leader determines who will be on the team and, in conjunction with the technical lead, what robots will be deployed. The team leader travels with the team to the field. During the field activity, the team leader continues coordination with authorities and serves as the "buck stops here" final decision maker. The team leader chairs the "hot wash" review after each field session and the overall after-action analysis at the end of the event. After the field activity, the team leader is responsible for any reports to authorities; that all equipment is refurbished and ready for the next event; and that issues identified in the after-action report are addressed.

• **Technical lead** The technical lead is the expert in either the types of robots or the mission. Prior to the field activity, the technical lead matches robots to the expected site conditions and logistical constraints. In the field, the technical lead allocates robots, generates worst-case scenarios for rehearsal preparation, and ensures all personnel know and apply correct operation and decontamination procedures.

• **Field lead** This person is the designated head of the squad of researchers and robots who are in the field for a specific mission. For example, a two-person squad drawn from a set of six team members may be deployed to an area. One of the members of the squad would be designated the field lead for that deployment in addition to his or her role as an operator, mission specialist, observer, and so forth. The principle is that although both members are responsible for safety, professional operation, and so forth, there is one person who will be the point of contact and who will make any final decisions if needed. Also, agencies expect a single name for the person in charge.

• **Team public information officer (PIO)** The team PIO is the person who interacts with the sponsoring agency PIO or serves as the direct liaison to the press. Before departing for the fieldwork, the team PIO determines the PIO chain of command, crafts the public information plan informing all team members as to what information can be released to whom (some information is private or would violate nondisclosure agreements), blogging policy, and so forth, and what the message is; for example, "These are prototype systems being graciously tested with agency A, and we are fortunate to be learning from them." Team members may be given a list of surprising or sensitive questions that should be answered with "Please ask person P as that is a bit beyond my scope." During the field activity, the PIO serves as the point of contact, ensures the public information plan is being enforced, prepares and edits any information to be used by the sponsoring PIO, and collates press. After the activity, the PIO distributes results, prepares thank-you notes, and performs other closure activities.

• **Robot expert operator** The robot expert operator is an expert in the operation, piloting, maintenance, and repair of the robot(s)—being able to pilot the robot is not enough. Before the field activity, the robot expert operator is responsible for checking out the robot, packing for transportation, and for making sure the technical lead is aware of logistical constraints (e.g., power, weather, etc.). During the activity, the robot expert operator will operate the vehicle or train others to use the robot and log the activities and performance. After the activity, the robot expert operator is responsible for field decontamination and for informing the technical lead of any problems, recommendation of changes in procedures, and so forth.

• **Mission specialist*** The mission specialist is the expert in the particular situation and represents the stakeholder; for example, the structural specialist determining whether a building is safe for entry, the hazardous materials specialist assessing the severity of the spill. The mission specialist is often a response professional that is assigned to the team at the time of the activity. The mission specialist directs the general operation, which the robot expert operator and field lead attempt to satisfy (though the field lead may override requests that put the robot at undue risk). The mission specialist is responsible for collecting the essential data needed for the actual or simulated mission.

• **Team safety officer** The team safety officer has overall responsibility for the safety of the team. Before the field activity, the team safety officer identifies any safety equipment and training needed for team members, vaccinations needed or health issues, and with the technical lead identifies and rectifies concerns in safe transportation of the robot (e.g., shipping limitations on batteries, internal combustion devices, etc.). During the field activity, the team safety officer makes sure each squad has a safety plan and a safety officer if needed and makes sure that improvements identified in the field are immediately incorporated and are captured in the after-action report. The team safety officer also is responsible for determining if a person is too fatigued to continue. Ideally, this role is not shared with the team leader or the technical lead roles so that there remains an independent viewpoint of safety to counter the tendency of the team leader and technical lead to focus on the expediency of the mission, but this may not be possible due to the need to minimize the number of people at the field event. The sponsoring agency generally provides safety oversight as well.

• **Field safety officer*** The field safety officer is the person who is responsible for safety during the squad operations in the field and can terminate operations, overriding the field leader if needed. The field safety officer is often a response professional that is assigned to the team at the time of the activity. During the rehearsal for the mission, the field safety officer postulates risks to the robots and personnel for the upcoming mission and makes sure the squad departs with safety gear, hydration, food, first aid kit, and any other relevant tools and supplies. During the mission, the field safety officer serves as the safety officer if needed (ground and marine operations generally do not require an oversight person, while aerial operations have a specific safety officer keeping watch over the UAV if it is flying over people and observing if the operator is distracted and about to make a mistake or trip and fall). The field safety officer can call for breaks or discontinue the activity. After the session, the field safety officer notes any possible improvements to safety to the team safety officer and makes sure these are brought up at "hot washes" so the improvements can be immediately incorporated and are captured in the after-action report.

• **Data manager** The data manager is responsible for preserving all documentation and data collected at the field activity, creating "snippets" for internal use by response professionals and for public release, and for ensuring privacy and security. Before the field activity, the data manager works with the team leader to create a data management plan capturing what will be recorded in the field, filename conventions, and backup procedures. During the activity, the first priority of the data manager is to gather all the incoming data, make sure it is labeled correctly, and perform backups. The data manager may also be in charge of recharging recording devices in order to take the load off of robot operators and scientific observers.

• **Scientific observer** Some field activities may have one or more trained scientific observers embedded with the squad to focus on opportunistic ethnographic observations or to conduct controlled experimentation. Before the field activity, the scientific observer works with the team leader to confirm that institutional review board approvals are in place if needed, to prepare recording devices and ensure sufficient redundancy in case of damage to or lost equipment in shipping or in the field, and to ensure sufficient recording media and mechanisms for recharging and backing up gear. During the field event, the scientific observer collects data. After the activity, the scientific observer works with the data manager to store and back up the data.

6.4.2   Selection of Robots and Gear

Selecting the best robots, bringing sufficient parts and tools, and managing the logistics of transportation and operations is critically important in fieldwork. There are four strategic considerations that influence the choice of robots for a field event; these are precursors to the tactical selection heuristics for specific robots within a modality as discussed in chapters 3–5.

What are the expected (and unarticulated) needs for robots? Having even an approximate understanding of the terrain or rubble types, characteristic dimensions of voids for ground searches, geographical extent of operations, and so forth, is helpful. But it is also important to remember that given the relatively small corpus of robot assistance at disasters and that because every disaster is different, what appears to be the best robot may not be the best fit in the field. Also, the need may change during the event; for example, at the Cologne Historical Archives building collapse response, the objectives changed from searching a pancake collapse of a multistory commercial building to the decidedly different rubble of a bricks-and-mortar house.

Diversity and redundancy of equipment and "system" thinking are the basic mechanisms for combatting surprises in the field. For example, a small UAV *system* is usually defined as two to three platforms, two redundant operator control stations, batteries and charger, spare parts, and the tools needed to effect field repairs. CRASAR generally brings two complete UGV systems. Squeezing in a small robot that doesn't take up space, such as the Access AC-ROV at the Tohoku tsunami response, can be a good idea. Sometimes, the only option is to accept the possibility that only one robot of a type can be brought, and if the robot fails and cannot be repaired, the field event will be over.

What are the transportation and logistical arrangements? The amount of gear and number of robots depends greatly on the transportation and logistical constraints. For example, robots that use internal combustion engines may not be transportable on airlines; likewise, some types of batteries may be forbidden as well. Controlled experimentation and exercises may have little constraints on what can be brought, but at a disaster, transportation, electricity, and even a secure place to leave gear while in the field are at a premium. Cases small enough to be acceptable by the airlines may be too large for personal cars and vans, which may be all that is available. Is a dedicated generator needed? Two? Or can everything be recharged from a car inverter?

How will maintenance and repairs be conducted? This question is both about bringing the right set of tools given space limitations and making sure there is space and lighting to work effectively. In deployments, it is important specifically to request workspace and power at the base of operations, such as a table for the data management gear and recharging plus a second table for working on robots. In field exercises, it is helpful to request a dedicated tent shelter, tables, and a cot for the data manager to nap in the lulls (and to serve as security). In other exercises, this might not be possible, so renting a suite where the living space can be converted to a workspace is often worth the added expense.

Are there any regulatory issues that must be considered (e.g., FAA, ITAR, FCC)? Unmanned aerial vehicles flying outdoors are generally regulated by the country in which they are flying, such as by the Federal Aviation Administration (FAA) in the United States. The Haitian national airspace was declared closed to unmanned vehicles in the aftermath of the 2009 earthquake, possibly in part to simplify air traffic control with manned helicopters operating at low altitudes. Note that in the United States, responders can request an emergency exemption to operate in a no-fly zone by applying for an **emergency COA**. Wireless transmitters may be on frequencies and wattages that fall under regulations such as by the Federal Communications Commission (FCC) in the United States or even if not regulated would impact other response operations. The use of robots or sensors in another country may be regulated by a country's customs laws or technology export rules, particularly the **International Traffic in Arms Regulations (ITAR)** in the United States.

6.4.3   Schedule in Field
The schedule for activities in the field depends largely on the type of field-work. In controlled experimentation, the researcher drives the design of the event and to some degree the exercises and concept experimentation. As a participant-observer, the robotics team adapts to the constraints of the requesting agencies.

The number of days in the field will vary, but *a rule of thumb is to limit the research team to 10 days in the field including travel*. If the event goes longer, then consider sending in a second, fresh team. This heuristic is a duplicate of the Federal Emergency Management Agency guidelines for deploying urban search and rescue teams. CRASAR experiences at the World Trade Center, Crandall Canyon Mine collapse, post—Hurricane Katrina UAV deployment, and the Tohoku tsunami response reinforce the rationale for this heuristic: at about the 10-day mark, people are so cumulatively tired and are working in such demanding conditions that cognitive fatigue becomes a continual problem. While there may be the temptation to stay longer for a controlled experiment because the conditions are less stressful, remember that just being in the field is stressful and even simple things like picking up parts in an unfamiliar town can take hours of frustration.

A typical day in a high-fidelity or real event would be a 12-hour shift, which can be longer due to transit and decontamination of the gear, combined with repair and recharging back at the base of operations. However, in general, *operators have only about 2 hours before they become noticeably cognitively fatigued*. Fortunately, in most events to date, the operating time for a run has been under 2 hours—mostly because the robots provide information quickly. As a result, operators in practice have historically gotten frequent breaks without the need to plan those in. But if a run goes longer, the field lead or the safety officer may need to call a break or swap operators. At the Crandall Canyon Mine collapse, the operators tried unsuccessfully for hours to withdraw the robot back into the borehole through the metal mesh lining the mine ceiling. After a night's sleep, the operation was performed the next day on the first try.

The *general flow of a shift has four phases. Before a squad departs for a shift* in the field, the team should discuss with the responders what are the objectives, constraints, limitations, safety, and what data they expect to get (and how). The safety officer should discuss the risk to the robot and what levels of risk are acceptable under which conditions. Any hasty training of responders should be conducted then. Just as each squad will have a single person in charge of the roboticists, if multiple responders embed with the robotics squad, the lead responder and the scope of his or her authority should be formally established. *Once in the field but before setting up for operations*, the field lead, the field safety officer, and the response agency representative should look over the site to decide what robots to use, a safe location for operations, and an egress plan in case of a sudden evacuation. *At the beginning of each run,* the field lead should follow a checklist to ensure everything is set up and the recording equipment is on and should hold a verbal "rehearsal" where the team talks about the specific mission, the strategies or steps, what can go wrong, and what each squad member should do in each case. Immediately after the squad returns, the people (especially boots and washing hands) and robots should be decontaminated, the data management process should be started, reports provided to the requesting agency, and logs updated.

Controlled experimentation typically has more team members involved, a shorter number of days in the field, and a shorter shift: My research group may field up to eight people for a controlled event. In many research events, we try to run two to three groups of responders at the same time because we have only a limited amount of time with them, for example for a day before or after a general training exercise. Each site would have a scientific observer in addition to the technical lead who was in charge of experiments overall, the team lead, a full-time data manager, and a dedicated person for repairing and recharging robots. The number of people in the field and the need to collect data according to the experimental method is a management challenge, and a practice session and written checklists and guidelines are often helpful. However, practice, guidelines, and written procedures may not be enough: The data from an entire 2-day exercise had to be discarded when one of the team members changed the search procedure during his shift to make it easier for the responders to use the robots. The loss of data set back research 6 months, and it was difficult to generate another $20,000 to cover the costs of another exercise and find another venue.

Exercises and concept experimentation may have less people and vary widely in the number of days and hours per shift. Agency-directed training exercises are of the order 3–5 days and may go longer than 12 hours so as to practice shift changes. The number and type of robots (and the need for redundancy) in the script of the exercise determines the number of people needed, often three to six. Concept experimentation such as the Summer Institutes that I started back in 2003 with National Science Foundation funding typically lasts 2–3 days, as the event usually identifies gaps between the robot system and the needs of the responders that can't be fixed overnight or the robots have broken and thus experimentation cannot continue.

6.5   Design of Data Collection

Regardless of the type of fieldwork, collecting data on the performance of the robot and human–robot system as a whole is paramount. This section describes six different categories of data to try to collect in the field: **a log of the activity, context, robot's-eye view, robot state, external view of the robot,** and **the human–robot interaction.** To collect these data reliably, it is helpful to create a data collection protocol. The design of the data collection protocol should take into account all of the constraints imposed by the circumstances of a particular fieldwork event, as some data collection sensors or methods may pose safety risks, may not function in the environmental conditions, or may be too intrusive into the operations. Any protocol for data collection must consider data management, or who will have access to what data after archiving.

6.5.1   Six Categories of Data to Collect

There are at least six types of data about the rescue robot and its operation within the larger system. It may not be possible to collect all of these data, but experience has shown that each of these is helpful. The six sources are listed in order of priority.

6.5.1.1   Log of Activity   A log of the activities is essential to establish the robot's performance and identify possible improvements. The log is usually handwritten by the robot team leader. Waterproof ink is desirable as are "smartpens" such as those made by Livescribe, which automatically generate PDFs of notes and can record audio.

At a minimum, the log of activity should capture

- where the robot was (GPS coordinates are useful)
- operating conditions and weather
- the start and stop time of the robot for each run
- who were the operators or team members and what were their roles
- what the mission was
- what the mission outcome was
- observations of failures or bottlenecks, and feedback from responders.

It is also useful to mark the time at which important events happened (e.g., that a worker's hard hat was found), so that a "snippet" or short clip of that can be created later. For example, an interesting structure or possible victim might appear in only 2 minutes of a 40-minute video from a robot run.

6.5.1.2 Context The context documents the overall operating conditions and what was going on. During an exercise or response, there is usually a plan of action for the day and maps in hardcopy that are useful to retain. A video of the site—both of where the robot is working and where the operators are—is also helpful.

We were unable to videotape or take pictures of the context at the World Trade Center collapse due to a prohibition intended to keep the press from intruding. This was particularly frustrating as the conditions on the rubble pile were quite different than those for which roboticists were designing mobility systems. Eventually, we realized that if we turned the video recorder for the robot on and then casually pointed the robot at the void and surroundings, we could obtain some contextual shots.

6.5.1.3 Robot's-Eye View Continually recording video of what the robot operator is seeing is critical to establishing what the robot was doing. Also as discussed in chapter 1, the responders will expect to replay and examine the recordings later even if they are present in the field. If a robot does not have onboard recording ability, taking a video of the screen is better than nothing.

6.5.1.4 Robot State Seeing what the robot was seeing is helpful, but to understand performance thoroughly, it is useful to have "black-box data"— data from all of the robot's sensors, what version of drivers were installed, what software programs were active, what commands or keystrokes the robot was given, network strength, and so forth. This allows analysts to reconstruct what inputs to the robot were given by the operator, what the robot received, what it sensed, and why it did what it did.

6.5.1.5   External View of the Robot   A video capturing what the robot was doing is also helpful in determining performance. This video may be captured by a second robot: As noted in chapter 3, it is common practice for a second UGV to follow the first in tight spaces or for tasks requiring mobile manipulation in order to give the operator a better view.

6.5.1.6   Human—Robot Interaction   A video of what the operating team was doing is useful for human—robot interaction analyses. This data can sometimes be captured by a safety officer or observer wearing a helmet camera. It should be noted that even if the data are captured during a disaster, the use of the data may need approval from an institutional review board.

6.5.2   Data Collection Protocol

The above is a significant amount of data to collect, with four of the six sources involving continual video recording. While the exact protocol for collecting and handling data depends on the robot, team, objectives, and type of fieldwork, it is important to have a clear protocol that everyone understands and has practiced before the event. The protocol should cover what you, the researcher, need but also what will be expected—the agencies or responders will want to see what you've collected immediately. This may mean dedicating a full-time person to editing and managing data.

Some suggestions on how to organize data collection are given in the following list.

Before the event:
- Create a protocol that specifies
    What data will be collected by whom.
    What media it will be stored on.
    The file-naming convention.
    How all the recorders will be synchronized with the right time/time zone/date.
    How often the data will be backed up and who will be the data manager. The data manager position is often a full-time job. During a deployment, it is helpful to have a single person who makes

sure all data are turned in and backed up. This person usually sleeps while the robot teams are in the field.

How the "chain of custody" will be managed and who will prepare snippets for agencies and other participants to look at.

- Host a practice session and identify how to simplify procedures or connections.
- Create "cheat sheets" and small tags attached to recorders or tools to remind researchers of the protocol.

During the event:

- After each run of the robot, while still in the field, confirm that critical data such as the robot's-eye view were recorded. This is so that if there was a recording failure, the mission can be rerun to get the data. If possible, note key snippets or tracks for priority analysis.
- After each session or day, have a formal outbrief with the team to (i) capture any safety or data collection protocol issues and (ii) record general observations and insights.
- After each session or day, back up all data and put that data in a secure location. At the World Trade Center collapse, the robot's-eye view data from the second day were lost when one group apparently took videotapes that had already been used for recording and were left out to be backed up. One of my students, Mark Micire, created a data management plan and checklist that was used for the remainder of the deployment. We positioned the tape decks by the front of the CRASAR area so that a returning group had to walk past the data manager on deck, who would remind the tired, hungry, and dirty operators to turn in their data.

At the end of the event but before leaving:

- Make sure that you have possession of all data from other agencies or sources. People get busy or distracted, and promised data often never materialize. We never received the video from the overhead camera of the robot working its way down the void at the Midas Gold Mine collapse.

6.5.3   Constraints on Data Collection Tools and Protocols

As would be expected, some data collection tools and protocols used for lab experiments are not necessarily appropriate nor will all work in a high-fidelity staged world or a natural world. Following are three questions a researcher should ask when preparing a data collection plan.

Does the tool or protocol affect safety? If the robot is operating in a sensitive environment, such as an explosive atmosphere, are the data collection tools intrinsically safe? Data cables or tethers that interfere with a robot's mobility or inhibit a human subject can be a problem. Likewise, a human subject would not be able to wear an eye-tracker in areas where safety glasses are needed. A major concern for an actual deployment is whether the data collection protocol requires more researchers in the hot zone, as minimizing the number of nonresponse personnel is critical for safety and for engendering acceptance of the robot.

Will the data gathering tool function in the environment or for the task? Video recorders may not be able to capture events in shadow. Eye-trackers may not work in situations where the test subject is not sitting at a desk and can walk around.

Is the tool intrusive? Sometimes there is not enough space in the work area to set up a tripod for recording or recording is viewed as intrusive. Biometric vests and other devices may constrict movement, add discomfort, or distract the human subject. Measurements such as the situation awareness global assessment technique (SAGAT) method for determining situation awareness require the operator to stop performing the task and take a test—this may not be acceptable for certain tasks or for high-fidelity missions.

6.5.4   Data Management

Collecting the data is not the same as archiving it, being able to find important elements later on, or being able to determine what can be shared with whom. Having a consistent file-naming convention and synchronization of data will help a great deal, but how best to organize and analyze field data is still an open question.

Some concerns about sharing data are as follows:

• If the data are from an exercise or deployment, the organizing agency should know up front what data it will get a copy of and what data will be released for scientific use. *All requests for use of the data for news or publicity should go through the agency's public information officer, and the agency should be acknowledged.*

• Human-subject data, including videos and pictures, have to be anonymized and *released in accordance with human-subject regulations*, managed by university **institutional review boards** in the United States.

• Data concerning victims, especially photographs and videos, are particularly sensitive and *typically are not released for general scientific use.*

• Data concerning a terrorist event or one that may involve lawsuits (such as a structural collapse) are also sensitive. Because these data may be used in legal proceedings, *it is imperative to maintain a "chain of custody" and be able to prove that the data are the original data.*

6.6 Participation in Responses

An actual response is a special case of fieldwork where the mission needs dictate the use of the robot, and data collection is secondary. It is also the case where the stresses of the actual operation pose additional constraints and the disturbing loss of life generates psychological risks. A major problem in actual responses is that the responders are overloaded and cannot adopt new technology (and the associated cognitive and logistics load), despite advances in robotics. Thus it is important to follow the basic etiquette developed by the CRASAR Roboticists Without Borders program. The psychological impact of participating in a response, particularly a large-scale event with a significant number of deaths, is a potential problem, and negative effects may manifest themselves sometime after the event.

6.6.1 Etiquette for a Response

While roboticists are convinced that their robots and expertise can make a difference, the ultimate decision for use of robots belongs to the professional responders, particularly the incident commander, and that decision must be respected. As professionals with legal accountability for the event, they have protocols and a collective history of expertise that follows a "keep it simple" strategy; unfamiliar and untrained people and robots do not help them keep the response simple and do not help them with what they have trained for.

Following is a list of observations about good etiquette.

• *Do not attempt first contact during a disaster* Responders also have a collective history of cases where people have volunteered to help but were not qualified. Likewise, untried tools require personnel to be diverted and can create unexpected consequences, invariably negative.

• *Do not self-deploy; you can be arrested, and Good Samaritan liability laws that protect responders do not apply* During the 2007 I-35 bridge collapse in Minneapolis, a group of researchers from Florida drove to Minneapolis and offered their advanced sonar equipment and services for a small fee. After being turned down, the team entered the site anyway, where they were arrested (Ingrassia, 2007).

• *Do not contact the families* During the 2007 Crandall Canyon Mine collapse response, a robotics hobbyist was so convinced that he could make a better robot than what had been used by the Mine Safety and Health Administration that he repeatedly contacted the families of the dead miners to pressure the MSHA to award him a grant to build the robot.

• *Do not expect to be reimbursed unless you have a prior agreement with the requesting agency*

• *Once at the site, accept "hurry up and wait"* While roboticists can see the value of immediate use, the responders have many facets of the disaster to handle and prioritize. During Hurricane Frances, I repeatedly left messages with Chief Ron Rogers, who was serving as incident commander for the state of Florida, asking when we would be deployed. Because the chief and I had known each other for many years, I forgot to consider that he was working at 150% capacity with his larger scope of responsibilities. My inquiries unforgivably distracted him.

• *Do not involve or bring the media* As noted earlier, the requesting agency will have a PIO, and everything must go through that person. Appearing to be more concerned about getting media attention than about helping with the response may cause the responders to not use the equipment and reflects poorly on the profession. In one CRASAR response, a team member invited a camera crew without clearing it with me or the requesting agency's PIO. The requesting agency chose not to give the crew any different access than the local media, which the camera crew repeatedly protested, annoying the responders and creating tension between the responders and CRASAR and me and the team member.

• *Follow the chain of command* Response is essentially organized along paramilitary lines. The point of contact is the person who gets the data first.

• *Acknowledge the requesting agency in all relevant publications and send copies.*

6.6.2 Psychological Safety

Rescue robotics is about disasters, destruction, human suffering, and loss of life. This can have significant psychological impact on roboticists, perhaps beyond what postevent counseling can mitigate. Therefore, it is imperative to select the field team judiciously and prepare the team members for what they might see and experience. Many of the CRASAR responders at the World Trade Center had active military experience. The students with me had been working and training with Florida Task Force 3, listening to their stories about disasters and seeing gruesome photographs. Although nothing can totally prepare a person for the magnitude of a disaster, the impact of what one sees (and smells) is lessened by preparation. During and after the response, it is the team leader's responsibility to observe the members, and himself or herself, for any signs of psychological distress. One unexpected consequence of being at a disaster is that the contrast between the disaster site and home is jarring. It is helpful to plan a transition period of a day before returning home so that the team members can discuss among themselves what they saw and felt.

6.7   Summary

There are four types of fieldwork: *controlled experimentation, participation in an exercise, concept experimentation,* and *participant-observer.* Controlled experimentation, participation in an exercise, and concept experimentation occur in a staged world, which is a high-fidelity, usually physical, simulation of one or more essential aspects of a disaster (e.g., physical rubble, night operations, entire team present, etc.). Participant-observer and opportunistic concept experimentation fieldwork are conducted at a disaster site. The roboticist has nearly total control of the activities, and there are few logistics constraints for controlled experimentation, whereas at the opposite end of the spectrum, the roboticist has little if any control of the activities and must operate under significant logistical constraints in a participant-observer event. Participating in a disaster response is at the request of the agency with incident command authority; roboticists self-deploy at the risk of arrest or distraction of the responders.

The fieldwork designer must plan the event, especially data collection. A field event will have six facets that affect the number of people and robots on the team and what can be accomplished: *fidelity, involvement of response professionals, logistics and tempo, risks, safety,* and *selection of robots and team members*. There are at least 10 distinct roles held by members of the team performing the fieldwork: *team leader, technical lead, field lead, team public information officer, robot expert operator, mission specialist, team safety officer, field safety officer, data manager,* and *scientific observer*. Individuals typically hold multiple roles, as a guiding principle is to minimize the number of people in the field. Robot operator effectiveness is generally around 2 hours, so a schedule must accommodate cognitive breaks.

A data collection plan with a formal protocol understood by all team members is essential. The data collection plan will usually include capturing a log of the activity, photographs or video of the site as context, video of the robot's-eye view, logs or snapshots of the robot state, video of the external view of the robot and what it was doing, and video of the human–robot interaction. Any protocol for data collection must consider data management, specifically who will have access to what data after archiving.

A good reference for designing fieldwork to understand and develop workable systems is *Code of Best Practice for Experimentation* by Alberts and Hayes (2002). The authors create a framework for visualizing what they call a campaign to develop systems that provide mission capabilities, originally for network-centric warfare, but this is applicable to disaster robotics as well. A campaign is a series of different types of fieldwork spanning the discovery of formative uses of the robots; the role of highly repeatable, but focused, test methods such as those of NIST in controlled, hypothesis-based experimentation; computer modeling and simulation; and demonstrations.

The trends discussed in chapter 2 show that robot deployments are increasing, but this does not mean researchers will participate in deployments and have more access to the field. The Fukushima Daiichi nuclear accident is an example where trained professionals or TEPCO workers used commercially available robots, with little involvement from developers and researchers. Exercises are relying more on use of robots from the agencies that have them and focusing on scripted demonstrations, which reduces the opportunity for researchers to participate and introduce advances. I see the role of CRASAR shifting from maintaining caches of robots and deployable teams to addressing the challenge of collecting data from these more "routinized" deployments and making it available to agencies, industry, and researchers. Responders will not have the time or the training to serve as participant-observers, organize any data collected, or to analyze performance; therefore, it is incumbent on researchers to create automated data collection and analysis tools, industry to incorporate them into its platforms, and agencies to insist on sharing data and conducting proper analyses.

Common Acronyms and Abbreviations

AUV   autonomous underwater vehicle; a UMV that operates underwater without a tether.

AUVSI   Association of Unmanned Vehicle Systems International.

COA   certificate of authorization; applies to UAVs in the United States.

COLREGS   International Regulations for Prevention of Collisions at Sea; maritime collision regulations that apply to UMVs.

COTS   commercial off-the-shelf; commercial technology that is readily available as opposed to custom built.

CRASAR   Center for Robot-Assisted Search and Rescue.

DARPA   Defense Advanced Research Projects Agency.

EO/IR   Electro-optical/infrared.

FAA   Federal Aviation Administration.

FCC   Federal Communications Commission.

FEMA   Federal Emergency Management Agency.

FLIR   Forward looking infrared.

IEEE   Institute of Electrical and Electronics Engineers.

IRS   International Rescue System Institute (Japan).

ITAR   International Traffic in Arms Regulations; many unmanned systems and sensors made in the United States fall under ITAR restrictions on exports.

MSHA   Mine Safety and Health Administration.

MAV   micro aerial vehicle.

NGO   nongovernmental organization.

NIOSH   National Institute for Occupational Safety and Health.

OCU   operator control unit.

OOTL   out of the loop control problem.

PIO   public information officer; a response agency or incident will have an assigned PIO.

ROV   remotely operated vehicle; connotes a tethered underwater vehicle.

RPA   remotely piloted aircraft.

RPV   remotely piloted vehicle; preferred by the U.S. Air Force over term *unmanned aerial vehicle* to emphasize the role of the human pilot.

sUAS   small unmanned aerial system; defined in the United States by the FAA as being under 55 pounds (25 kilograms) and working at limited altitudes.

TTX   tabletop exercise.

UAS   unmanned aerial system; preferred by FAA over term *unmanned aerial vehicle* in order to emphasize the role of the human pilot in control and the integration into the national airspace.

UASI   Urban Area Security Initiative; a U.S. grant program that enables public safety agencies to buy and share technology.

UAV   unmanned aerial vehicle.

UGV   unmanned ground vehicle.

UMV   unmanned marine vehicle.

US&R   urban search and rescue; FEMA abbreviation.

USAR   urban search and rescue; common abbreviation.

USA TACOM–ARDEC–EOD   U.S. Army Tank Automotive and Armaments Command–Army Research and Development Command–Explosive Ordnance Disposal Division.

USV   unmanned surface vehicle; a UMV that operates on the surface of a body of water.

UUV   unmanned underwater vehicle; the set of all types of unmanned systems that operate below the surface of the water.

UxV   the universal set of unmanned vehicles, where "x" is replaced by A (aerial), G (ground), or M (marine) based on context.

WTC   World Trade Center (New York).

Glossary

accessibility elements A term taken from architecture that refers to doors, stairs, ladders, or any component of the habitable space that controls the choice of movement.

accident An unintended action(s) that results in a disaster such as a building, bridge, or tunnel collapse, a chemical emergency, dam failure, hazardous materials release, nuclear event, or train wreck.

adaptive shoring Using robots dynamically to shore unstable rubble to expedite the extrication process.

authenticity How well a simulation represents what happens in the natural world.

autonomous capabilities Cognitive functions delegated to the robot beyond platform stabilization (e.g., navigation, health maintenance).

biomimetic robot A robot whose design replicates some aspect of a biological creature; generally connotes the design of the physical platform, as most artificial intelligence software for robotic control is biologically inspired.

cold zone The region outside of the working area of a disaster where the responders set up their support activities.

concept experimentation Experimentation conducted in a staged world or natural world that is focused on generating concepts of how a new technology or protocol can be used, leading to a mission capability package.

controlled experimentation Experimentation conducted in a staged world, as opposed to a laboratory setting, that is highly focused on a single topic or attribute.

data collection protocol Specifies what will be collected and how it will be collected reliably, safely, and with minimal intrusion into operations.

data manager The person who is responsible for preserving all documentation and data collected at the field activity, for creating "snippets" for internal use by response professionals and for public release, and for ensuring privacy and security.

deep water   Regions outside the littoral region; depths much greater than 60 meters.

deployability   How people get the robots to the point of ingress for the mission.

direct intervention   When a UxV directly interacts with the environment to effect mitigation of disaster such as by manipulating valves or emergency devices.

disaster   An event that requires resources beyond the capability of a community and necessitates a multiple-agency response.

disaster life cycle   Historically conceptualized as four phases consisting of *prevention, preparation, response*, and *recovery*. Other life cycles that emphasize that these four activities may (or should) occur in parallel exist.

disaster robotics   The study of any modality of unmanned system that may apply to any or all of the phases of a disaster life cycle.

egocentric view   The view relative to the agent's internal frame of reference; a camera provides an egocentric view.

egress   The point of exit.

emergency COA   A certificate of authorization (COA) granted by the FAA in a matter of hours in order to allow a UAV to be used at a disaster.

entombed victim   A victim beneath the surface of a collapse who is not crushed or severely pinned by the debris.

estimation of debris volume and types   A major mission during the recovery phase that accelerates cleanup and enable residents to reenter affected areas.

ethnography   A set of techniques for studying human systems and the resulting description.

exercise   A success driven event in a staged world designed to maximize role playing or demonstration of the favorable contributions of the technology.

exocentric view   A representation of a scene from an external frame of reference; for example, looking at the robot as opposed to looking through the robot.

extreme incident   An event that is handled internally within a fire rescue department by a team with specialized training, somewhat similar to a police action by a special weapons and tactics (SWAT) team.

field lead   The person who is the designated head of the squad of researchers and robots who are in the field for a specific mission.

field safety officer   The person who is responsible for safety during the squad operations in the field and can terminate operations, overriding the field leader if needed.

fixed-based serpentine robots   A snake robot attached to another robot like an elephant trunk.

fixed-wing craft   Aircraft with wings like a plane.

forensics   A mission where robots deliver structural sensor payloads to more favorable viewing angles to determine the cause of a disaster (e.g., structural forensics) or gather evidence (e.g., photograph a crime scene).

formative work domain   A term taken from industrial organization psychology for a work domain where a technology or protocol is creating a new capability rather than providing an alternative method for an existing function.

free serpentine robots   Snake robots that are free-standing and locomote through direct contact with a surface.

geological events   Earthquakes, tsunamis, and volcanoes.

guarded motion   Where the robot protects itself by limiting or modifying human directives that would cause a collision or other problem.

hasty training   Connotes on-site, rapid training of response professionals for use of robots at a disaster.

holistic evaluation   Evaluation of a robot in a larger sociotechnical system, where performance includes how well it fits the constraints of the existing workflow.

hot zone   The region restricted due to the hazard; the actual disaster area.

human–robot interaction   The consideration of the entire sociotechnical system, including questioning the choice of what responsibilities are handled by the human or the robot (cognitive science), how the human teams with the robot (psychology, communications), as well as the human–computer interface design (computer science).

human scale   Refers to a site large enough and still intact enough that a human outfitted in protective gear may walk through the site; thus, the site does not limit the size of a robot that may be used.

ingress   Point of entry.

in situ medical assessment and intervention   A mission for a robot to permit medical personnel to triage victims at the point of injury and to provide life support functions for the victim such as transporting and administering fluids and medication.

interior operations   Operations within a manmade structure, such as a building or mine, or natural, such as a cave, or other spatial regions not considered outside.

littoral   Regions of shallow water near shore, generally extending from the high-tide mark to about 60 meters from shore.

local-area operations   Operations near a single structure or terrain feature or a small cluster of structures or features; in contrast to wide-area operations.

logistics support   A mission that automates the transportation of equipment and supplies from storage areas to teams or distribution points within the hot zone.

man-packable   A robot small enough to be carried to the point of ingress in a backpack; if a UGV, this connotes it can work in areas of interest too small for humans and dogs.

man-portable   A robot that can be carried by two people; if a UGV, this connotes it can work in the same scale as a small human but can be carried to the point of ingress.

maxi   A robot that needs a trailer or cannot be lifted and carried by responders; if a UGV, this connotes it can work in exteriors and spaces that a human could maneuver through, but it is heavy and cumbersome.

medically sensitive extrication and evacuation of casualties   A mission where UGVs may be needed to help provide medical assistance while victims are still in the disaster area, also known as the hot zone.

meteorological events   Avalanches, floods, fires, heat waves, hurricanes, thunderstorms, tornadoes, wildfires, and winter storms.

micro UAV   A UAV that is less than 1 meter in any characteristic dimension and weighs less than 1 kilogram. Alternative definition: a man-packable UAV that fits completely in one backpack.

micro UGV   A man-packable robot that fits completely in one backpack.

mini UGV   A man-packable robot that can be split and carried in two backpacks.

mining- and mineral-related disasters   These are caused by either a meteorological event or an accident; typically these are managed by the owner, not the government.

mission capability package   How a new technology or protocol can be used; it includes the technology but also the training requirements and protocols to be used by the decision-making hierarchy.

mission specialist   The person who is the expert in the particular situation and represents the stakeholder.

mobile beacon or repeater   A device to extend wireless communication coverage and bandwidth and to localize personnel based on their radio signal strength.

mobile robots   The term used synonymously in robotics and automation for *unmanned ground vehicles* or in artificial intelligence for any physically situated robot.

modality of a robot   Whether it is a ground, aerial, or marine vehicle.

motes   Robots so small that their mobility depends on the wind or other influences in the environment.

natural world   The unconstrained real world; for example, an actual disaster as opposed to an exercise.

negative obstacles   Hindrance to navigation below the plane of travel; these are more commonly referred to as "holes" or slopes or stairs heading down.

non-navigational constraint   Specifies whether a robot can be used for a unique environmental factor; for example, is the robot intrinsically safe for use in explosive atmospheres?

normative work domain   A term taken from industrial organization psychology for a work domain where a technology or protocol provides an alternative mechanism for an existing function performed by a human, usually by duplicating the existing actions and procedures.

operational envelope   The nominal environment where the robot operates, which may be composed of regions with different environmental constraints.

operational morality   The ethical responsibility of a manufacturer or operator to ensure that the robot will perform reliably and to make users aware of implicit consequences of use.

operator's environment   Where the OCU is located and operators or responders viewing and directing the robot are working.

organic   The robot is deployed, maintained, transported, and tasked and directed by the stakeholder

participant-observer   A person who collects ethnographic data about a team or event while serving as a team member, rather than purely observing.

physically situated agent   A person, robot, or software entity that can sense and act in an environment.

pilot primary viewpoint   Refers to where the pilot is looking during operations; one of the two factors in classifying the style of control of a UAV.

polycentric   A system where there are multiple ("poly") places or organizations ("centers").

positive obstacles   Obstacles above the plane of travel.

preparedness phase   The phase in the disaster life cycle devoted to preparing for a disaster.

prevention phase   The phase in the disaster life cycle devoted to prevention of disaster; often, a robot can function in prevention as well as response and recovery (e.g., a port inspection robot).

quantitative measurability   The ability to extract quantifiable measurements, either from empirical observations or from controlled experimentation.

reconnaissance and mapping   A mission where the robot provides responders with general situation awareness and an understanding of the destroyed environment.

recovery phase   The phase in the disaster life cycle devoted to reentry of citizens into the affected area and economic recovery.

remote presence   A human sensing and acting at a distance in real time; teleoperation is the least sophisticated mechanism for this.

repeatability   The ability to duplicate the key aspects of a situation or event; disasters are not repeatable due to the variations in the event, personnel, and interdependencies.

rescue robotics   Robotics applied to the postevent phases of a disaster, usually the response phase and sometimes the recovery phase.

response phase   The phase in the disaster life cycle devoted to saving lives and mitigating the disaster; this is often called search and rescue, though more activities occur.

robot expert operator   The person who is an expert in the operation, piloting, maintenance, and repair of the robot(s).

robot system   Refers to everything needed to use a robot (e.g., the robot plus OCU and any necessary batteries, fuel, etc.) rather than referring to the robot platform.

rotorcraft or rotary-wing craft   Helicopter types of aircraft.

rubble removal   A mission where robotic machinery or exoskeletons assist with either extrication of victims or rebuilding.

scale of the region   The metric that determines if the robot can fit and is likely to maneuver successfully in a region; for example, is the robot small enough to fit in the tunnel?

scientific observer   A person who is embedded with the squad to focus on opportunistic ethnographic observations or to conduct controlled experimentation.

scientific reductionism   Decomposing a problem into narrow hypotheses that can be tested with statistically significant, reproducible experimentation.

search   Aims at finding a victim and any potential hazards in extrication.

sense and avoid   A term used by the aviation industry for how aircraft avoid midair collisions with other aircraft; for UAVs this included collisions with obstacles.

sensor fouling   Degradation of a sensor due to becoming covered or clogged with dirt, water, mud, or other environmental factors; once fouled, moving out of a region does not clear the sensor.

severity   The number and size of obstacles in the robot's operational envelope.

staged world   A physical simulation that replicates a set of key features of the natural world with high fidelity.

strategic asset   A satellite, a military or Coast Guard helicopter, or a high-altitude drone, which are deployed and tasked by an agency that then passes information to tactical teams as it becomes available.

structural inspection   A mission that can be facilitated through robots that deliver structural sensor payloads to more favorable viewing angles.

surface victims   People lying on the surface of the rubble or readily visible to a rescuer.

surrogate   When a robot serves as the physical embodiment of a remote team member, such as when team members outside the hot zone use telepresence to work with team members in the hot zone to monitor the state of progress and anticipate needs.

survivability constraints   The effects of the environment on the failure rate of the robot.

tabletop exercise (TTX)   One type of staged world, generally focuses on roles and interactions between echelons in a command hierarchy.

tactical   An asset that is directly controlled by stakeholders with "boots on the ground"—people who need to make fairly rapid decisions about the event.

tactical, organic asset   A system generally small enough and portable enough to be transported, used, and operated on demand by the group needing the information; also called a tactical, organic system.

team leader   Person who is the overall leader of an entire group conducting fieldwork.

team public information officer   Person who interacts with the sponsoring agency PIO or serves as the direct liaison to the press.

team safety officer   The person who has overall responsibility for the safety of the team.

technical lead   The person who is the expert in either the type of robots or the mission.

terrorism   An action(s) intended to destroy man-made facilities such as buildings, bridges, tunnels, ports, and critical infrastructure.

tortuosity A term taken from fractals and ethology meaning the amount of twistiness or number of turns in a space.

trapped victim A victim beneath the surface of a collapse, either immobilized or partially crushed.

traversability of a region The conditions within the region that affect navigation; for example, can the robot drive through the mud and over rubble in the tunnel?

unattended ground sensor Often a mote or a robot that has moved into a desirable location to provide sensing.

unmanned systems A term for ground, aerial, and marine robots to distinguish them from robots used for factory automation.

verticality Slope of the terrain being traversed.

victim recovery A mission where the robot assists in extricating bodies already found or in finding all of the missing.

warm zone The restricted area around the hot zone where responders enter and exit the hot zone and decontaminate equipment and people.

wide area A geographically distributed work area, such as for an earthquake or hurricane.

References

Alberts, D. S., and R. E. Hayes. 2002. *Code of Best Practice for Experimentation*. CCRP Publication Series. Department of Defense (USA) Command and Control Research Program. Washington, DC: U.S. Department of Defense.

Ambrosia, V. G., et al. 2010. The Ikhana unmanned airborne system (UAS) western states fire imaging missions: From concept to reality (2006–2010). *Geocarto International* 26 (2):85–101.

Andersson, K. P., and E. Ostrom. 2008. Analyzing decentralized resource regimes from a polycentric perspective. *Policy Sciences* 41:71–93.

Angermann, M., M. Frassl, and M. Lichtenstern. 2012. Mission review of aerial robotic assessment—ammunition explosion Cyprus 2011. In *IEEE International Symposium on Safety, Security and Rescue Robotics*, pp. 1–6. College Station, TX.

Arai, M., et al. 2008. Development of "Souryu-IV" and "Souryu-V:" Serially connected crawler vehicles for in-rubble searching operations. *Journal of Field Robotics* 25 (1–2):31–65.

Barbera, J. A., C. DeAtley, and A. G. Macintyre. 1995. Medical aspects of urban search and rescue. *Fire Engineering* 148:88–92.

Bethel, C., and R. Murphy. 2010. Non-facial and non-verbal affective expression for appearance-constrained robots used in victim management. Paladyn. *Journal of Behavioral Robotics* 1 (4):219–230.

Blanchard, B. W. 2007. *Guide to Emergency Management and Related Terms, Definitions, Concepts, Acronyms, Organizations, Programs, Guidance, Executive Orders & Legislation*. Emmitsburg, MD: FEMA Emergency Management Institute. Available at: http://training.fema.gov/EMIWeb/edu/termdef.asp.

Blitch, J. G. 1996. *Knobsar: An Expert System Prototype for Robot Assisted Search and Rescue*. M.S. thesis, Department of Applied Mathematics and Statistics, Colorado School of Mines, Golden, CO.

Blitch, J., R. R. Murphy, and T. Durkin. 2002. Mobile semiautonomous robots for urban search and rescue. In *Encyclopedia of Microcomputers*, Vol. 28, ed. A. Kent and J. G. Williams, pp. 211–224. New York: Marcel Dekker.

Bob99Z. 2010. ROV's collide. Available at: www.youtube.com/watch?v=iQq_pPj fWmQ.

Boyd, J. 2011. IEEE Blogs Automation: Robotic construction machine causes explosion at Fukushima. Available at: http://spectrum.ieee.org/automaton/robotics/industrial-robots/robotic-construction-machine-causes-explosion-at-fukushima.

Burke, J. L. 2006. *RSVP: An Investigation of the Effects of Remote Shared Visual Presence on Team Process and Performance in US&R Teams*. Ph.D. thesis, Department of Industrial Organization Psychology, University of South Florida, Tampa.

Burke, J. L., and R. Murphy. 2004. Human-robot interaction in USAR technical search: Two heads are better than one. In *13th IEEE International Workshop on Robot and Human Interactive Communication, 2004 (ROMAN 2004)*, pp. 307–312.

Burke, J., and R. Murphy. 2007. RSVP: An investigation of remote shared visual presence as common ground for human-robot teams. In *Proceedings of the Second ACM SIGCHI/SIGART Conference on Human-Robot Interaction (HRI 2007)*, pp. 161–168. Arlington, VA.

Burke, J., et al. 2004. Moonlight in Miami: An ethnographic study of human-robot interaction in USAR. *Human-Computer Interaction* 19 (1–2):85–116.

Carlson, J., and R. R. Murphy. 2005. How UGVs physically fail in the field. *IEEE Transactions on Robotics* 21 (3):423–437.

Carlson, J., R. Murphy, and A. Nelson. 2004. Follow-up analysis of mobile robot failures. In *2004 IEEE International Conference on Robotics and Automation (ICRA)*, Vol. 5, pp. 4987–4994.

Casper, J., and R. R. Murphy. 2000. Issues in intelligent robots for search and rescue. In *SPIE Unmanned Ground Vehicle Technology II, 2000*, pp. 292–302. Orlando, FL.

Casper, J., and R. Murphy. 2002. Workflow study on human-robot interaction in USAR. In *IEEE International Conference on Robotics and Automation (ICRA 2002)*, pp. 1997–2003.

Casper, J., and R. R. Murphy. 2003. Human-robot interaction during the robot-assisted urban search and rescue effort at the World Trade Center. *IEEE Transactions on Systems, Man and Cybernetics, Part B* 33 (3):367–385.

Chang, C., and R. R. Murphy. 2007. Towards robot-assisted mass-casualty triage. In *IEEE International Conference on Networking, Sensing and Control (2007)*, pp. 267–272.

Chazan, G. (2010, 17 May). Spill fight shows progress. *Wall Street Journal*. http://online.wsj.com/article/SB10001424052748704614204575245810666495600.html.

Cosenzo, K. A., R. Parasuraman, A. Novak, and M. Barnes. 2006. Implementation of automation for control of robotic systems (Report ARL-TR-3808). Aberdeen Proving Grounds, MD: Army Research Laboratory.

Craighead, J., B. Day, and R. Murphy. 2006. Evaluation of Canesta's range sensor technology for urban search and rescue and robot navigation. IEEE Workshop on Safety Security Rescue Robots.

Davids, A. 2002. Urban search and rescue robots: From tragedy to technology. *IEEE Intelligent Systems* 17 (2):81–83.

Duncan, B. A., and R. R. Murphy. 2012. Methods and metrics of autonomous take-off, landing, and GPS waypoint navigation experiments in micro-UAVs. In *International Conference on Unmanned Aircraft Systems*, pp. 1–12.

Duncan, B., and R. R. Murphy. 2013. Field study identifying barriers and delays in data-to-decision with small unmanned aerial systems. In *International Conference on Technologies for Homeland Security*.

Endsley, M. 1988. Design and evaluation for situation awareness enhancement. In *Human Factors Society 32nd Annual Meeting*, Vol. 32, pp. 97–101.

FBI. 2007. Photo gallery: FBI response to Minneapolis bridge collapse. Available at: www.fbi.gov/news/stories/2007/august/minn2_photos082107.

FEMA. 2013. Learn about the types of disasters. Available at: www.ready.gov/natural-disasters.

Ferreira, F. 2011. *IEEE Robotics and Automation Society Newsletter* ICRA Special Forum: "Preliminary Report on the Disaster and Robotics in Japan." Available at: http://wiki.ieee-ras.org/mab/newsletterjune2011report1.

Fincannon, T., et al. 2004. Evidence of the Need for Social Intelligence in Rescue Robots. In *IEEE/RSJ International Conference on Intelligent Robots and Systems, 2004. (IROS 2004)*, Vol. 2, pp. 1089–1095.

Gage, A., R. R. Murphy, and B. Minten. 2013. Shadowbowl 2003: Lessons learned from a reach-back exercise with rescue robots. *IEEE Robotics & Automation Magazine* 11(3):62–69.

Goldwert, L. 2007. Minneapolis honors bridge collapse victims. Available at: www.cbsnews.com/stories/2007/08/08/national/main3146827.shtml?tag=mncol;lst;1.

Goodrich, M. A., et al. 2008. Supporting wilderness search and rescue using a camera-equipped mini UAV. *Journal of Field Robotics* 25 (1–2):89–110.

Humphrey, C. M. 2009. Information Abstraction Visualization for Human-Robot Interaction. Ph.D. thesis, School of Engineering, Vanderbilt University, Nashville, TN.

Humphrey, C. M., and J. Adams. 2009. Robotic tasks for CBRNE incident response. *Advanced Robotics* 23 (9):1217–1232.

Ingrassia, R. 2007. Man arrested at bridge only trying to help, lawyer says. Broadcast on August 9, 2007, WISN ABC 12 Minneapolis.

Kaber, D. B., and M. R. Endsley. 1997. Out-of-the-loop performance problems and the use of intermediate levels of automation for improved control system functioning and safety. *Process Safety Progress* 16 (3):126–131.

Kawatsuma, S., M. Fukushima, and T. Okada. 2012. Emergency response by robots to Fukushima-Daiichi accident: Summary and lessons learned. *Industrial Robot International Journal (Toronto, Ont.)* 39 (5):428–435.

Kobayashi, A., and K. Nakamura. 1983. Rescue robot for fire hazards. In *Proceedings of the 1983 International Conference on Advanced Robotics*, pp. 91–98. Tokyo, Japan.

Krotkov, E., and J. Blitch. 1999. The Defense Advanced Research Project Agency (DARPA) Tactical Mobile Robotics program. *International Journal of Robotics Research* 18 (7):769–776.

Kruijff, G.-J., et al. 2012. Rescue robots at earthquake-hit Mirandola, Italy: A field Report. In *IEEE International Symposium on Safety, Security and Rescue Robotics*, pp. 1–8. College Station, TX.

Leitsinger, M. 2011. Tsunami town's fishermen vow to "bring joy back." Japan after the wave. Available at: http://worldblog.nbcnews.com/_news/2011/06/17/6879693-tsunami-towns-fishermen-vow-to-bring-joy-back.

Lester, J., A. Brown, and J. Ingham. 2012. Christchurch Cathedral of the Blessed Sacrament: Lessons learnt on the stabilisation of a significant heritage building. In *New Zealand Society for Earthquake Engineering (NZSEE) Annual Conference*, pp. 1–12.

Linder, T., et al. 2010. Rescue robots at the collapse of the Municipal Archive of Cologne City: A field report. In *IEEE International Workshop on Safety, Security, and Rescue Robotics (SSRR 2010)*, pp. 1–6.

Low, V. 2010. Robot to detect survivors in flooded mine. Malaysian National News Agency. Available at http//www.bernama.com/bernama/v5/newsgeneral.php?id=487247.

Lussier, C. 2010. MDSU 2 clears the way for humanitarian relief. Military Sealift Command Press Release. Available at http//www.msc.navy.mil/N00p/pressrel/press10/press08.htm.

Manzi, C., M. J. Powers, and K. Zetterlund. 2002. *Critical Information Flows in the Alfred P. Murrah Building Bombing*. Washington, DC: Chemical and Biological Arms Control Institute.

McClean, D., ed. 2010. *World Disasters Report 2010: Focus on Urban Risk*. Geneva: International Federation of Red Cross and Red Crescent Societies.

McClure, M., D. R. Corbett, and D. W. Gage. 2009. The DARPA LANdroids program. In *SPIE Unmanned Systems Technology XI*, ed. G. R. Gerhart, D. W. Gage, and C. M. Shoemaker, p. 7332A. Bellingham, WA.

Michael, N., et al. 2012. Collaborative mapping of an earthquake-damaged building via ground and aerial robots. *Journal of Field Robotics* 29 (5):832–841.

Micire, M. 2002. *Analysis of Robotic Platforms Used at the World Trade Center Disaster*. M.S. thesis, Department of Computer Science and Engineering, University of South Florida.

Micire, M. J. 2008. Evolution and field performance of a rescue robot. *Journal of Field Robotics* 25 (1–2):17–30.

Murphy, R. R. 2000. Marsupial and shape-shifting robots for urban search and rescue. *IEEE Intelligent Systems* 15 (3):14–19.

Murphy, R. R. 2002. Rats, robots, and rescue. *IEEE Intelligent Systems* 17 (5):7–9.

Murphy, R. 2003. Rescue robots at the World Trade Center. *Journal of the Japan Society of Mechanical Engineers* 102 (1019):794–802 [special issue on Disaster Robotics].

Murphy, R. R. 2004a. Human-robot interaction in rescue robotics. *IEEE Transactions on Systems, Man and Cybernetics, Part C: Applications and Reviews* 34 (2):138–153.

Murphy, R. 2004b. Trial by fire. *IEEE Robotics & Automation Magazine* 11 (3):50–61.

Murphy, R. R. 2006. Fixed- and rotary-wing UAVs at Hurricane Katrina. In *IEEE International Conference on Robotics and Automation* (video proceedings).

Murphy, R. R. 2010a. Navigational and mission usability in rescue robots. *Journal of the Robotics Society of Japan* 28:142–146.

Murphy, R. R. 2010b. Potential uses and criteria for unmanned aerial systems for wildland firefighting. In *AUVSI Unmanned Systems North America* (CD only). Arlington, VA: AUVSI.

Murphy, R. R. 2011a. The 100:100 challenge for computing in rescue robotics. In *IEEE International Symposium on Safety Security Rescue Robots*, pp. 72–75.

Murphy, R. R. 2011b. Use of unmanned systems for search and rescue 2010. In *AUVSI North America* (CD only). Arlington, VA: AUVSI.

Murphy, R., and J. L. Burke. 2008. From remote tool to shared roles. *IEEE Robotics and Automation Magazine* 15 (4):39–49.

Murphy, R. R., and J. L. Burke. 2010. The safe human-robot ratio. In *Human-Robot Interactions in Future Military Operations*, ed. F. J. M. Barnes, pp. 31–49. Burlington, VT: Ashgate.

Murphy, R. R., and J. Shields. 2012. *The Role of Autonomy in DoD Systems*. Defense Science Board Task Force Report. Washington, DC: Department of Defense.

Murphy, R. R., and R. Shoureshi eds. 2008. *Emerging Mining Communication and Mine Rescue Technologies*. Pittsburgh: Mine Safety and Health Administration. 419 pp.

Murphy, R. R., and S. Stover. 2006a. Gaps analysis for rescue robots. In *ANS 2006: Sharing Solutions for Emergencies and Hazardous Environments*. Salt Lake City, UT.

Murphy, R. R., and S. Stover. 2006b. Rescue robot performance at 2005 La Conchita mudslides. In *ANS 2006: Sharing Solutions for Emergencies and Hazardous Environments*. Salt Lake City, UT.

Murphy, R. R., and S. Stover. 2008. Rescue robots for mudslides: A descriptive study of the 2005 La Conchita mudslide response: Field reports. *Journal of Field Robotics* 25 (1–2):3–16.

Murphy, R. R., and D. D. Woods. 2009. Beyond Asimov: The three laws of responsible robotics. *IEEE Intelligent Systems* 24 (4):14–20.

Murphy, R. R., J. Burke, and S. Stover. 2006. Field studies of safety security rescue technologies through training and response activities. *IEEE International Symposium on Safety, Security, and Rescue Robotics (SSRR 2006)*. Gaithersburg, MD. (CD only)

Murphy, R., K. Pratt, and J. L. Burke. 2008. Crew roles and operational protocols for rotary-wing micro-UAVs in close urban environments. In *2008 3rd ACM/IEEE International Conference on Human-Robot Interaction (HRI)*, pp. 73–80. Amsterdam.

Murphy, R. R., D. Riddle, and E. Rasmussen. 2004. Robot-assisted medical reachback: A survey of how medical personnel expect to interact with rescue robots. In *13th IEEE International Workshop on Robot and Human Interactive Communication (ROMAN 2004)*. pp. 301–306.

Murphy, R., S. Stover, and H. Choset. 2005. Lessons learned on the uses of unmanned vehicles from the 2004 Florida hurricane season. In *AUVSI Unmanned Systems North America* (CD only). Arlington, VA AUVSI.

Murphy, R., J. Casper, J. Hyams, M. Micire, and B. Minten. 2000. Mobility and sensing demands in USAR. In *26th Annual Conference of the IEEE Industrial Electronics Society, 2000 (IECON 2000)*, pp. 138–142.

Murphy, R. R., S. Stover, K. Pratt, and C. Griffin. 2006a. Cooperative damage inspection with unmanned surface vehicle and micro aerial vehicle at Hurricane Wilma. In *IEEE/RSJ International Conference on Intelligent Robots and Systems (IROS 2006)*, pp. 1–9. Bejing, China.

Murphy, R. R., T. Vestgaarden, H. Huang, and S. Saigal. 2006b. smart lift/shore agents for adaptive shoring of collapse structures: A feasibility study. In *IEEE Workshop on Safety Security Rescue Robots (SSRR 2006)*. Gaithersburg, MD.

Murphy, R. R., et al. 2006c. Use of micro air vehicles at Hurricane Katrina. In *IEEE Workshop on Safety Security Rescue Robots (SSRR 2006)*. Gaithersburg, MD.

Murphy, R. R., et al. 2006d. Use of unmanned surface and aerial vehicles to inspect damage after Hurricane Wilma. In *AUVSI Unmanned Systems North America* (CD only). Arlington, VA: AUVSI.

Murphy, R., et al. 2008a. Cooperative use of unmanned sea surface and micro aerial vehicle at Hurricane Wilma. *Journal of Field Robotics* 25 (3):164–180.

Murphy, R. R., et al. 2008b. Rescue robotics. In *Handbook of Robotics*, ed. B. Sciliano and O. Khatib, pp. 1151–1174. New York: Springer-Verlag.

Murphy, R., et al. 2009a. Mobile robots in mine rescue and recovery. *IEEE Robotics & Automation Magazine* 16 (2):91–103.

Murphy, R. R., et al. 2009b. Requirements for wildland firefighting ground robots. In *AUVSI North America* (CD only). Arlington, VA: AUSVI.

Murphy, R. R., E. Steimle, M. Lindemuth, D. Trejo, M. Hall, D. Slocum, S. Hurlebas, and Z. Medina-Cetina. 2009c. Robot-assisted bridge inspection after Hurricane Ike. In *IEEE Workshop on Safety Security Rescue Robotics*, pp. 1–5. Denver, CO: IEEE.

Murphy, R. R., A. Rice, N. Rashidi, Z. Henkel, and V. Srinivasan, 2011a. A multi-disciplinary design process for affective robots: Case study of Survivor Buddy 2.0. In *2011 IEEE International Conference on Robotics and Automation*, pp. 701–706. Shanghai, China.

Murphy, R. R., et al. 2011b. Robot-assisted bridge inspection. *Journal of Intelligent & Robotic Systems* 64 (1):77–95.

Murphy, R. R., et al. 2012a. Marine heterogeneous multi-robot systems at the Great Eastern Japan Tsunami recovery. *Journal of Field Robotics* 29 (5):819–831 [special issue on Heterogeneous Multiple Robots].

Murphy, R., et al. 2012b. Projected needs for robot-assisted chemical, biological, radiological, or nuclear (CBRN) incidents. In *10th IEEE International Symposium on Safety, Security, and Rescue Robotics (SSRR 2012)*, pp. 1–4. College Station, TX.

Nagatani, K., et al. 2011. Redesign of rescue mobile robot Quince. In *IEEE International Symposium on Safety, Security, and Rescue Robotics (SSRR 2010)*, pp. 13–18. Kyoto, Japan.

Nagatani, K., et al. 2013. Emergency response to the nuclear accident at the Fukushima Daiichi Nuclear Power Plants using mobile rescue robots. *Journal of Field Robotics* 30 (1):44–63.

National Fire Protection Association. 1999. *Standard on Operations and Training for Technical Rescue Incidents*. Avon, MA: National Fire Protection Association.

Newman, P. 2010. Beneath the horizon. *Engineering & Technology* 5 (12):36–39.

NewsCore. 2011. Underwater robots inspect Japanese port. Fox News Atlanta. http://www.myfoxphilly.com/dpps/news/underwater-robots-inspect-japanese-port-dpgonc-20110421-ch_12867062 (retrieved 21 April 2011).

Norman, J. 2012. *Fire Officer's Handbook of Tactics*. 4th ed. Fire Engineering.

Ohno, K., et al. 2011. Robotic control vehicle for measuring radiation in Fukushima Daiichi Nuclear Power Plant. In *2011 IEEE International Symposium on Safety, Security, and Rescue Robotics (SSRR 2011)*, pp. 38–43. Kyoto, Japan.

Onosato, M., et al. 2012. Digital Gareki archives: An approach to know more about collapsed houses for supporting search and rescue activities. In *IEEE International Symposium on Safety, Security, and Rescue Robotics (SSRR 2012)*, pp. 1–6. College Station, TX.

Peschel, J. M. 2012a. *Mission Specialist Human-Robot Interaction in Micro Unmanned Aerial Systems*. Ph.D. thesis, Department of Computer Science and Engineering, Texas A&M University: College Station, TX.

Peschel, J. M. 2012b. Towards physical object manipulation by small unmanned aerial systems. In *IEEE International Symposium on Safety, Security and Rescue Robotics*, pp. 1–6. College Station, TX.

Peschel, J., and R. R. Murphy. 2013. On the human-computer interaction of unmanned aerial system mission specialists. *IEEE Transactions on Human-Machine Systems* 43:53–62.

Platt, D. 2002. The use of robots as a search tool. *Fire Engineering* 155 (10). Available at: http://www.fireengineering.com/articles/print/volume-155/issue-10/world-trade-center-disaster/volume-ii-the-ruins-and-the-rebirth/the-use-of-robots-as-a-search-tool.html.

Pratt, K. S. 2007. *Analysis of VTOL MAV Use during Rescue and Recovery Operations Following Hurricane Katrina*. M.S. thesis, Department of Computer Science and Engineering. University of South Florida.

Pratt, K., and R. R. Murphy. 2012. Protection from human error: Guarded motion methodologies for mobile robots. *IEEE Robotics & Automation Magazine* 19 (4):36–47.

Pratt, K., et al. 2006. Requirements for semi-autonomous flight in miniature UAVs for structural inspection. In *AUVSI Unmanned Systems North America* (CD only). Arlington, VA: AUVSI.

Pratt, K., R. Murphy, J. Burke, J. Craighead, C. Griffin, and S. Stover. 2008. Use of tethered small unmanned aerial system at Berkman Plaza II collapse. In *IEEE International Workshop on Safety, Security, and Rescue Robotics*, pp. 134–130.

Pratt, K., et al. 2009. CONOPS and autonomy recommendations for VTOL SUASs based on Hurricane Katrina operations. *Journal of Field Robotics* 26 (8):636–650.

Riddle, D. R., R. Murphy, and J. L. Burke. 2005. Robot-assisted medical reachback: Using shared visual information. In *IEEE International Workshop on Robot and Human Interactive Communication (ROMAN 2005)*, pp. 635–642.

Rizzo, A. 2010. Search ends for missing U.S. balloonists. *Washington Times*. http://www.washingtontimes.com/news/2010/oct/4/search-ends-missing-us-balloonists/ (retrieved 4 October 2010).

Rogers, E. M. 2003. *Diffusion of Innovations*. 5th ed. New York: Free Press.

Science Daily. 2006. Wasps: Man's new best friend! Entomologists train insects to act like sniffing dogs. Available at: www.sciencedaily.com/videos/2006/0702-wasps_mans_new_best_friend.htm.

Shibuya, M. 2012. Using micro-ROV's in the aftermath of Japan's tsunami. In *Underwater Intervention 2012* (CD only). New Orleans, LA.

Sizemore, J. 2010. *Small UAS (sUAS) Special Federal Aviation Rule (SFAR) Part 107*. Washington, DC: Federal Aviation Administration.

Smith, M. R. 2012. EMILY lifeguard robot is the next wave in water rescues. *Huffington Post*. Available at: http://www.huffingtonpost.com/2012/08/17/emily-lifeguard-robot_n_1794067.html.

Srivaree-Ratana, P. 2012. Lessons learned from the Great Thailand Flood 2011: How a UAV helped scientists with emergency response and disaster aversion. In *AUVSI Unmanned Systems North America* (CD only). Arlington, VA: AUVSI.

Statement Under Oath of Virgil Brown. 2011. In *Upper Big Branch Accident Investigation Interviews*. Mine Safety and Health Administration, www.msha.gov/PerformanceCoal/Transcripts/I-0042.pdf.

Strickland, E. 2011. 24 hours at Fukushima. *IEEE Spectrum* 48 (11):35–42.

Tadokoro, S., R. Murphy, S. Stover, W. Brack, M. Konyo, T. Nishimura, and O. Tanimoto. 2009. Application of active scope camera to forensic investigation of construction accident. In *IEEE International Workshop on Advanced Robotics and Its Social Impacts (ARSO2009)* pp. 47–50.

Tittle, J. S., A. Roesler, and D. D. Woods. 2002. The remote vision problem. In *Proceedings of the 46th Meeting of the Human Factors & Ergonomics Society*, pp. 260–264.

TVNZ. 2010a. 24 hours of obstacles slows rescue progress. Available at: http://tvnz.co.nz/national-news/24-hours-obstacles-slows-rescue-progress-3908163.

TVNZ. 2010b Pike River mine explosion: Day 5 as it happened. Available at: http://tvnz.co.nz/national-news/pike-river-mine-explosion-day-5-happened-3907934.

TVNZ. 2010c. Pike River mine explosion: Day 6 as it happened. Available at: http://tvnz.co.nz/national-news/pike-river-mine-explosion-day-6-happened-3912855.

United States Fire Administration. 1996. *Technical Rescue Program Development Manual*. Emmitsburg, MD: United States Fire Administration.

Vicente, K. J. 1999. *Cognitive Work Analysis: Toward Safe, Productive, and Healthy Computer-Based Work*. Mahwah, NJ: Lawrence Erlbaum.

Voyles, R. M., and A. C. Larson. 2005. TerminatorBot: A novel robot with dual-use mechanism for locomotion and manipulation. *IEEE/ASME Transactions on Mechatronics* 10 (1):17–25.

Wikipedia. 2012. Remotely operated underwater vehicle. http://en.wikipedia.org/wiki/Remotely_operated_underwater_vehicle.

Woods, D., and E. Hollnagel. 2006. Joint Cognitive Systems: Patterns in Cognitive Systems Engineering. Boca Raton, FL: CRC Press, Taylor & Francis.

Index

Accidents, types, 1, 14, 52, 58, 60, 63, 87, 109

Authenticity, role in research, 163, 164, 165, 166, 169, 173

Biomimetic robot types, 66

Center for Robot-Assisted Search and Rescue (CRASAR), 2, 4, 18, 25, 28, 29, 30, 31, 32, 36, 37, 39, 40, 47, 48, 61, 80, 81, 82, 104, 126, 130, 134, 144, 152, 154, 155, 156, 158, 170, 172, 173, 177, 179, 181, 185, 186, 191, 193, 194, 196

Certificate of authorization (COA), 129 emergency COA, 186

Cooperative perception, 87, 147

Data to collect, categories, 188–190

Defense Advanced Research Projects Agency (DARPA), 1, 3, 18, 25, 58, 66, 71, 77, 84, 85, 104, 113

Disasters
Alfred P. Murrah Federal Building bombing, USA, 1995 (also known as Oklahoma City bombing), 2, 3, 16, 71
Barrick Gold Dee Mine event, USA, 2002, 24, 25, 84, 88
Berkman Plaza II collapse, 2007, 6, 7, 23, 24, 31, 33, 44, 56, 74, 79, 85, 90, 92, 93, 119, 120, 123, 124, 130

Browns Fork Mine event, USA, 2004, 24, 26

California wildfires, 2006, 6, 116

Chernobyl nuclear accident, Soviet Union, 1986, 3

Christchurch earthquake, New Zealand, 2011, 24, 36, 91, 117, 118, 119, 120, 121, 134

Crandall Canyon Mine disaster, USA, 2007, 23, 24, 31, 32, 40, 43, 44, 46, 50, 54, 69, 77, 78, 79, 90, 92, 93, 94, 95, 96, 104, 178, 186, 187, 193

Deepwater Horizon explosion, USA, 2010 (also known as BP Oil spill or Macondo blowout), 11, 14, 24, 34, 41, 45, 48, 53, 54, 141, 146, 151, 153

DR No. 1 Mine event, USA, 2005, 24, 27, 47, 84, 89, 93, 94, 97

Excel No. 3 Mine event, USA, 2004, 24, 27, 44, 80, 84, 89, 93

Finale Emilia earthquake, Italy, 2012, 24, 38, 58, 119, 120, 121, 123, 130, 133, 134

Fukushima Daiichi nuclear accident, Japan, 2011, 4, 6, 11, 15, 17, 22, 24, 37–38, 42, 45, 46, 47, 48, 50, 54, 55, 56, 58, 66, 69, 71, 76, 77, 84, 85, 87, 91, 92, 93, 94, 95, 96, 103, 104, 105, 106, 109, 116, 119, 120, 122, 123, 129, 131, 133, 154, 196

Disasters (cont.)

Great Thailand flood, Thailand, 2011 (also known as 2011 Thailand monsoon flooding), 23, 24, 38, 116, 120, 129, 134

Disasters

Haiti earthquake, Haiti, 2010, 23, 24, 33, 52, 56, 118, 119, 120, 129, 146, 149, 150, 186

Hanshin-Awajii (Kobe) earthquake, Japan, 1995, 2

Historical Archive of Cologne building collapse, Germany, 2009, 24, 39, 72, 79, 80, 87, 90

Hurricane Charley, USA, 2004, 24, 39, 80–82, 85, 87, 89, 109

Hurricane Ike, USA, 2008, 24, 32, 45, 48, 55, 56, 137, 144, 146, 150, 152, 153, 172

Hurricane Katrina, USA, 2005, 4, 7, 24, 29, 53, 56, 80, 81, 82, 89, 109, 116, 117, 118, 119, 120, 121, 122, 123, 125, 127, 129, 130, 134, 176, 186

Hurricane Wilma, USA, 2005, 24, 29, 30, 56, 120, 123, 130, 133, 145, 146, 147, 150, 152

I-35 bridge collapse, USA, 2007, 24, 31, 35, 145, 146, 150, 153, 193

Jim Walter No. 5 Mine explosion, USA, 2001, 24, 39, 69, 87, 88

La Conchita mudslide, USA, 2005, 6, 24, 28, 30, 44, 46, 49, 75, 89, 93, 94

L'Aquila earthquake, Italy, 2009, 24, 33, 34, 36

McClane Canyon Mine, USA, 2005, 24, 28, 44, 47, 84, 89, 93

Mexico City earthquake, Mexico, 1985, 10

Midas Gold Mine collapse, USA, 2007, 24, 30, 31, 41, 50, 70, 72, 80, 89, 93, 97, 98, 177, 191

Missing balloonists, Adriatic Sea, 2010, 23, 24, 35, 141, 151

Niigata Chuetsu earthquake, Japan, 2004, 24, 26, 27, 28, 88, 104

Pike River Mine explosion, New Zealand, 2010, 6, 13, 23, 24, 36, 41, 43, 45, 46, 48, 54, 56, 69, 91, 92, 93, 94, 95, 178

Prospect Towers collapse, USA, 2010, 13, 14, 24, 35, 79, 85, 90

Sago Mine disaster, USA, 2006, 6, 24, 30, 44, 48, 73, 89, 93, 95

San Juan De Sabinas Coal Mine explosion, Mexico, 2011, 40

San Jose Copper—Gold Mine, Chile, 2010, 40, 54

Three Mile Island nuclear accident, USA, 1979, 3

Tohoku earthquake, Japan, 2011, 15, 24, 36, 58, 104, 117, 118, 120, 130, 133

Tohoku tsunami, Japan, 2011, 7, 10, 15, 24, 36, 45, 47, 58, 91, 137, 142, 146, 147, 149, 151, 152, 153, 156–158, 179, 185, 186

Upper Big Branch Mine explosion, USA, 2010, 13, 14, 24, 39, 40, 87, 90

Wangjialing Coal Mine flooding, China, 2010, 13, 24, 39, 69, 88, 90

World Trade Center collapse, USA, 2001 (also as known as 9/11 collapse), 2, 3, 4, 6, 9, 11, 22, 23, 24, 25–26, 28, 44, 47, 48, 50, 55, 56, 58, 69, 70, 71, 77, 80, 88, 92, 93, 94, 95, 96, 99, 102, 105, 173, 174, 176, 177, 179, 186, 189, 191, 194

Disaster life cycle

preparedness phase, 1, 13, 18, 59, 117, 161

prevention phase, 1, 13, 18, 59, 117, 161

recovery phase, 1, 12, 13, 15, 16, 18, 23, 25, 29, 31, 37, 38, 42, 59, 63, 71, 84, 92, 93, 111, 117, 119, 137, 139, 142, 146, 154, 159, 161 response phase, 1, 12, 13, 16, 18, 22, 56, 63, 71, 111, 118, 137, 141, 142, 159, 161, 175, 176, 186, 189, 193

Economic justification for disaster robots, 2, 9, 10, 13, 18
Experimentation
  concept, 163, 164, 167, 168, 169, 171–172, 173, 179, 186, 188, 195
  controlled, 163, 164, 166, 168–169, 170, 173, 184, 185, 186, 187, 195, 196
  exercises, 163, 164, 169–170, 185, 186, 188, 196
  participant-observer, 163, 164, 165, 168, 169–171, 173, 174, 186, 195, 196
Extreme operating conditions, 5, 6, 56, 125, 143, 160, 189
Extreme terrains, 5, 6, 8, 94, 105, 107, 164, 178

Failure
  due to breakage or malfunction, 42
  due to external factor or environment, 42
  due to human error, 42
  nonterminal, 49
  rates, UGVs, 41
  taxonomy, 43
  terminal, 43–49
Florida Task Force 3, 29, 39, 82, 83, 99
Formative work domains, 15, 19, 59, 60, 171, 196

Geological disaster events, 14, 16, 17, 18, 58, 59, 116
  avalanches, 14
  earthquakes, 1, 2, 13, 14, 15, 23, 24, 26, 27, 28, 33, 34, 36, 38, 53, 56, 58, 60, 63, 80, 87, 88, 91, 104, 116, 117, 118, 119, 120, 121, 123, 129, 130, 133, 134, 146, 150, 151, 186

landslides, 6, 14, 24, 28, 30, 44, 46, 49, 52, 60, 75, 87, 89, 93, 94
tsunamis, 7, 10, 14, 15, 24, 36, 45, 47, 58, 137, 142, 146, 147, 149, 152, 153, 154, 156, 157, 179, 185, 186
volcanoes, 14
GPS- and wireless-denied environments, 5, 7, 55, 56, 63, 86, 132, 143, 160
Guarded motion or guarded autonomy, 110, 111, 114, 129, 132, 134, 135

Head-up, head-down operation, 114
Holistic evaluation, 163, 164, 166, 167
Human–robot interaction, 5, 8, 9, 49, 59, 60, 83, 99, 105, 107, 108, 109, 122, 129, 131, 132, 135
Human scale of operational spaces, 3, 53, 83, 101, 102, 108

Indiana Task Force 1, 179
Institutional review boards, 192
International Rescue System Institute (IRS), 26, 28, 31, 33, 36, 37, 38, 39, 40, 46, 47, 154, 155, 156

Japanese Ground Self-Defense Forces, 37

Keyhole effect, 8

Localize, observe general surroundings, look specifically for victims, report (LOVR) protocol, 97–98

Manipulation, 34, 42, 52, 63, 64, 107, 109, 147, 190
  categories: investigatory and intervention, 87, 147
Man-made disaster events, 14
  chemical emergencies, 11, 14, 16, 55, 67, 71, 116, 117, 128, 131, 135, 169, 170
  dam failures, 14

Man-made disaster events (cont.)
  nuclear incidents, 3, 4, 11, 14, 15, 16,
    17, 22, 24, 37, 38, 46, 47, 52, 55, 58,
    60, 66, 68, 69, 84, 92, 103, 109, 116,
    120, 123, 128, 154, 167, 196
  structural collapse, 3, 4, 6, 7, 8, 9, 11,
    14, 16, 17, 19, 23, 24, 25, 31, 32, 33,
    35, 36, 39, 40, 43, 47, 48, 50, 52, 53,
    54, 55, 56, 57, 58, 59, 60, 66, 69, 70,
    71, 72, 74, 75, 79, 80, 82, 83, 84, 85,
    87, 92, 93, 94, 102, 103, 105, 107,
    109, 117, 119, 123, 125, 130, 133,
    140, 144, 145, 146, 150, 152, 153,
    172, 173, 174, 177, 178, 179, 180,
    185, 186, 187, 189, 191, 192, 193
  train wrecks, 14
Marine environment, littoral or deep
  water, 141
Meteorological disaster events, 14, 16,
    17, 18, 58, 59, 109, 116
  fires, 14, 86
  Meteorological disaster events
  floods, 14, 23, 24, 38, 39, 116, 120,
    129, 133
  heat waves, 14
  hurricanes, 4, 14, 24, 29, 32, 39, 45,
    52, 53, 55, 56, 60, 80–82, 85, 87, 89,
    109, 116, 117, 119, 120, 123, 125,
    127, 129, 130, 133, 134, 137, 144,
    145, 146, 147, 150, 152, 153, 166,
    172, 176
  tornadoes, 14
  winter storms, 14
Mining or mineral related disaster
  events, 14, 15, 18, 39
Mission environment plot, 101
Missions, list, 15–17
  acting as a mobile beacon or repeater,
    16, 18, 59, 71, 72, 85, 95
  adaptive shoring, 16, 17, 59, 85
  direct intervention, 10, 17, 19, 34, 42,
    52, 59, 63, 64, 84, 87, 107, 108, 109,
    145, 146, 147, 161, 190

  estimation of debris volume and
    types, 17, 18, 52, 59, 146, 149, 150–
    151, 161
  in situ medical assessment and inter-
    vention, 16, 59, 86, 109
  medically sensitive extrication and
    evacuation of casualties, 16, 86, 109
  providing logistics support, 17, 18, 59, 85
  reconnaissance and mapping, 13, 15,
    18, 23, 36, 38, 52, 56, 59, 83, 84, 95,
    107, 109, 117, 118, 120, 129, 132,
    133, 134, 135, 145, 146, 147, 157,
    159, 161, 169
  rubble removal, 16, 17, 85, 109, 146
  search, 2, 15, 18, 23, 25, 28, 29, 31,
    35, 39, 40, 47, 51, 52, 54, 56, 59, 72,
    74, 79, 81, 82, 83, 84, 88–91, 92, 93,
    94, 105, 109, 141, 145, 146, 147,
    149, 150–151, 152, 154, 158, 161
  serving as a surrogate for a team mem-
    ber, 17, 18, 59, 85, 86, 105
  structural inspection and forensics, 13,
    15, 16, 18, 29, 31, 32, 36, 59, 60, 72,
    84, 85, 88–91, 116, 118, 120, 134,
    145, 146, 149, 150–151, 161, 177
  victim recovery, 17, 18, 31, 59, 83, 84,
    86, 88–91, 109, 145, 146, 150–151,
    154, 161

National Aeronautics and Space Administration, 3, 6, 8, 85, 97
New Jersey Task Force, 1, 12, 14, 18, 35,
  53, 60
Non-navigational constraints, 68
  maintainability, 71
  sensing, 69
  types: survivability, 69
  unintended consequence, 71

Performance, 8, 41–51
  standards, 58, 60
  UAV, 119–122
  UGV, 92–97

UMV, 149–154
Physically situated agency, 4, 73, 74,
    101, 102, 103
Point of ingress or egress, 68, 77–80
    accessibility element, 77
    engineered breach or borehole, 77
    natural void, 77

Quantitative measurability, 163, 164,
    165

Remote presence, 2, 8, 10, 56, 100, 108,
    110, 138, 139, 140, 141, 160
Repeatability, need for, 107, 163, 164,
    165, 166
RoboCup Rescue League, 3
Robots used at disasters or related
    events
    AC-ROV, Access, 37, 149, 151, 155, 185
    Active Scope Camera UGV, Interna-
        tional Rescue System Institute, 31,
        33, 39, 54, 67, 80, 90
    AEOS 1 USV, AEOS, 29, 30, 140, 150, 152
    ANDROS Wolverine UGV, Remotec, 4,
        23, 25, 26, 27, 28, 30, 39, 40, 66
    BROKK90, BROKK330D, or
        BROKK800D UGV, Brokk, 37, 91
    custom ROV, Tokyo Institute of Tech-
        nology, 37
    custom ROV, Tokyo University, 37
    custom UAV, Ascending Technologies,
        33, 34, 56
    custom UAV, Natural Human–Robot
        Cooperation in Dynamic Environ-
        ments Project (NIFTi), 38, 117, 118,
        120
    custom UAV, Siam UAV Industries, 38,
        116, 120, 129
    custom UGV, Natural Human–Robot
        Cooperation in Dynamic Environ-
        ments Project (NIFTi), 38
    custom UGV, Western Australia Water
        Company, 36, 91

Falcon UAV, Ascending Technologies,
    38
Global Hawk UAV, Northrop Grum-
    man, 4, 6, 111, 116
Hummingbird UAV, Ascending Tech-
    nologies, 38
Ikhana UAV, NASA, 6, 116
IP-3 UAV, iSENSYS, 7, 31, 44, 47, 56,
    120, 122, 127, 133
Kenaf UGV, International Rescue Sys-
    tem Institute, 36, 91
KOHGA3 UGV, International Rescue
    System Institute, 36, 91, 104
Little Benthic Vehicle (LBV) ROV, Sea-
    Botix, 33, 35, 150, 155
Maxximum ROV, Oceaneering, 34
Micro-Tracks or VersaTrax UGV, Inuk-
    tun, 4, 25, 35, 65, 88, 102
Micro-VGTV or VGTV UGV, Inuktun,
    4, 25, 28, 31, 32, 39, 44, 46, 65, 82,
    88, 89, 94, 105
Millenium ROV, Oceaneering, 34
Mine Cavern Crawler UGV, Inuktun
    Mine, 31, 32, 40, 46, 77, 78, 90, 93,
    95, 104
Mitsui ROV, Mitsui, 37
New Zealand Defense Force bomb
    squad robot UGV, unknown manu-
    facturer, 36, 43, 45, 46, 47, 91, 95
Oceanmapper or Echomapper AUV,
    YSI, 32, 37, 45, 47, 138, 139, 157
Packbot UGV, iRobot, 3, 4, 16, 37, 38,
    45, 46, 47, 66, 72, 76, 85, 88, 91, 92,
    94, 95, 103, 106
Parrot AR.drone UAV, 36, 117, 120,
    121
Pelican UAV, Ascending Technologies,
    36, 117, 120
Predator UAV, General Atomics, 4,
    111, 116, 118
Quince UGV, International Rescue Sys-
    tem Institute, 36, 38, 45, 46, 47, 54,
    91, 93, 94, 95, 104

Robots used at disasters or related
  events (cont.)
  Raven UAV, AeroVironment, 29, 56,
    114, 118, 120, 121, 129, 130
  RESQ-A and custom UGVs, Japan
    Atomic Energy Agency, 38, 38, 91,
    104
  SARbot ROV, SeaBotix, 7, 37, 48, 138,
    147, 151, 152, 155, 157
  Sea-RAI USV, AEOS, 32, 56, 138, 140,
    144, 150
  Skylark UAV, Elbit, 34, 118, 120, 129
  Solem, QinetiQ (also known as Foster-
    Miller), 25, 44, 47, 69, 70, 88, 93, 94,
    104, 177
  Souryu UGV, International Rescue Sys-
    tem Institute, 26, 27, 28, 88, 104
  Talon UGV, QinetiQ (also known as
    Foster-Miller), 3, 4, 25, 37, 38, 44,
    47, 48, 66, 67, 88, 91, 94, 95, 96,
    103, 106
  T-Hawk UAV, Honeywell, 4, 37, 41,
    116, 119, 120, 121, 123, 127, 129
  T-Rex UAV, Like90 or iSENSYS, 29
  unknown UAV, Japan Air Photo Ser-
    vice Company, 37
  Ultra Heavy Duty ROV, Schilling, 34
Robots used at disasters or related
  events
  VGTV Xtreme or Extreme UGV, Inuk-
    tun or American Standard Robotics,
    25, 28, 29, 30, 35, 39, 44, 49, 65, 67,
    70, 75, 83, 89, 90, 94, 97, 177
  Video Ray ROV, Video Ray, 31, 32, 37,
    45, 48, 144, 150, 154
  Warrior UGV, iRobot, 38, 91
Roles in fieldwork, 181–184

Scale of a region, 68, 73–74
  exterior spaces, 73
  granular space, 74
  habitable space, 73
  ratio, 73

  restricted maneuverability space, 73
Sense and avoid, 114
Sensor fouling, 44, 47, 50, 94, 95, 110
Situation awareness, 15, 42, 51, 99, 110,
  116, 121, 132, 160, 161, 192
Staged world simulations, 164, 165,
  166, 167, 168, 169, 171, 176, 191,
  195
Style of control UAV, 113
  autonomous capabilities, 113
  pilot primary viewpoint, 113
Swift-water rescue, 14

Tabletop exercises (TTXs), 51, 163, 167
Teleoperation, 8, 113
Traversability, 74–77
Trench collapse, 14

U.S. Coast Guard, 4, 12, 14, 35, 130, 158
U.S. Federal Emergency Management
  Agency (FEMA), 9, 12, 14, 53, 179
U.S. Mine Safety and Health Adminis-
  tration (MSHA), 4, 12, 14, 18, 23, 24,
  26, 27, 28, 30, 31, 36, 39, 40, 47, 48,
  49, 53, 60, 71, 72, 92, 104, 194
U.S. National Guard, 4, 13, 118

Visual navigation, 132

Wide-area operational environments,
  115
Wilderness search and rescue, 14, 115,
  116
Workarounds, 49
  human adaptation, 50
  physical or software kludges, 50

Intelligent Robotics and Autonomous Agents

Edited by Ronald C. Arkin

Dorigo, Marco, and Marco Colombetti, *Robot Shaping: An Experiment in Behavior Engineering*

Arkin, Ronald C., *Behavior-Based Robotics*

Stone, Peter, *Layered Learning in Multiagent Systems: A Winning Approach to Robotic Soccer*

Wooldridge, Michael, *Reasoning About Rational Agents*

Murphy, Robin R., *An Introduction to AI Robotics*

Mason, Matthew T., *Mechanics of Robotic Manipulation*

Kraus, Sarit, *Strategic Negotiation in Multiagent Environments*

Nolfi, Stefano, and Dario Floreano, *Evolutionary Robotics: The Biology, Intelligence, and Technology of Self-Organizing Machines*

Siegwart, Roland, and Illah R. Nourbakhsh, *Introduction to Autonomous Mobile Robots*

Breazeal, Cynthia L., *Designing Sociable Robots*

Bekey, George A., *Autonomous Robots: From Biological Inspiration to Implementation and Control*

Choset, Howie, Kevin M. Lynch, Seth Hutchinson, George Kantor, Wolfram Burgard, Lydia E. Kavraki, and Sebastian Thrun, *Principles of Robot Motion: Theory, Algorithms, and Implementations*

Thrun, Sebastian, Wolfram Burgard, and Dieter Fox, *Probabilistic Robotics*

Mataric, Maja J., *The Robotics Primer*

Wellman, Michael P., Amy Greenwald, and Peter Stone, *Autonomous Bidding Agents: Strategies and Lessons from the Trading Agent Competition*

Floreano, Dario and Claudio Mattiussi, *Bio-Inspired Artificial Intelligence: Theories, Methods, and Technologies*

Sterling, Leon S., and Kuldar Taveter, *The Art of Agent-Oriented Modeling*

Stoy, Kasper, David Brandt, and David J. Christensen, *An Introduction to Self-Reconfigurable Robots*

Lin, Patrick, Keith Abney, and George A. Bekey, editors, *Robot Ethics: The Ethical and Social Implications of Robotics*

Weiss, Gerhard, editor, *Multiagent Systems, second edition*

Vargas, Patricia A., Ezequiel A. Di Paolo, Inman Harvey, and Phil Husbands, editors, *The Horizons of Evolutionary Robotics*

Murphy, Robin R., *Disaster Robotics*